W9-AAW-539

Homestead Year

Back to the Land in Suburbia

For Ted (Migrant Worker)
and
to the Memory of Anna and Harlan Hubbard
and their homestead in Payne Hollow

Printed in the United States of America

Design by Catherine Lau Hunt
Illustrations by Molly Burk, page 45 and 201.
Illustrations by Leslie Watkins, pages 13, 32, 40, 58, 77, 101, 105, 189, 215, 234, and 269.

10 9 8 7 6 5 4 3 2 1

Library of Congress Cataloging-in-Publication Data

Moffett, Judith, 1942–
    Homestead year : back to the land in suburbia / Judith Moffett.
    p. cm.
    ISBN 1-55821-352-X
    1. Organic farming—Pennsylvania—Rose Valley.   2. Moffett, Judith, 1942—
Homes and haunts—Pennsylvania—Rose Valley.   3. Urban agriculture—
Pennsylvania—Rose Valley.   I. Title.
S605.5.M625   1995
635'.0484'097481—dc20   95-2997
                                                                            CIP

Parts of this book have appeared, in somewhat different form, in the following publications: *Country Journal, The Green Scene: The Magazine of the Pennsylvania Horticultural Society,* and *The Pennsylvania Gazette: Alumni Magazine of the University of Pennsylvania.*

Introduction

Homesteading. The back-to-the land "movement" of the Sixties and Seventies gave to the word, and the concept, a charge of moral fervor. Back then, twenty or thirty years ago, the idea of trying to achieve economic independence from mainstream society by embracing a lifestyle close to nature, which put the important things first, had enormous appeal for a lot of young (and not-so-young) people who moved to the country in couples or in groups and tried, with widely varying degrees of success and persistence, to establish themselves there. Helen and Scott Nearing's description of their own venture into self-sufficiency in New England—*Living the Good Life,* originally published in 1954—was reissued in 1970. *Payne Hollow: Life on the Fringe of Society,* by Harlan Hubbard, appeared from Eakins Press in 1974. And during the Seventies, Rodale Press published dozens of philosophical and how-to books about homesteading, emphasizing the organic approach characteristic of the Nearings and the Hubbards, and consistent with the classic homesteading value of independence from the chemicals and fossil-fuel consumption which were widely viewed as devastating to the environment.

Then came the Eighties, Reaganomics, and the yuppification of society. Suddenly young people were no longer interested in trying to live off the land; the combination of virtue and hard physical work that had made this an attractive option no longer had appeal. Making life choices out of concern for social issues was out of fashion. Campuses turned Republican. Doing without material things looked plain idiotic to the new crop of college graduates bound for Wall Street and lucrative careers in medicine and law. And Rodale Press—a bellwether of trends in matters of this sort—gradually stopped publishing books about raising goats and building root cellars and passive solar green-houses for growing vegetables intensively twelve months a year; by 1990 such titles had almost disappeared even from its backlist. Instead, the Rodale catalog featured books about health, fitness, nutrition, cooking, landscaping, woodworking, and stress control. Rodale was still committed to gardening and sustainable agriculture, and was still emphatically organic, but the emphasis had shifted: instead of *The Homesteader's Handbook to Raising Small Livestock,* the fall 1990 catalog of-fers *The Homeowner's Complete Handbook for Add-On Solar Greenhouses and Sunspaces.* (Note that word "Homeowner" and all that it implies.) Among the nearly three hundred titles listed were a few homesteader-type backlist books on raising poultry and rabbits, preserving food, even Maurice Telleen's classic *The Draft Horse Primer* (all listed under "Farming"); but the audience for the five-acres-and-independence-type books which had been such a strength of Rodale's list during the Seventies had apparently dried up.*

In the fall of 1984, just as interest in homesteading was giving way to interest in things for which making money is essential, my own interest in living a homesteader's life was springing unaccountably into flower. I was forty-two years old, married to a professor of English at the University of Pennsylvania, and in my own seventh year as an assistant professor of English at Penn. We lived in Delaware County outside Philadelphia, in a small house with a heavily wooded lot, not very suit-able for gardening. I had published two books of poetry with university presses, one of Swedish translations, and one of literary criticism; and after some years of teaching a science-fiction course, I was becoming very interested in trying to write some science fiction myself. That was the chief reason why I had already decided not to go ahead with the tenure-review process, but to switch to part-time teaching.

*The best source of such works at present is *Storey's How-To Books for Country Living* cat-alog, descriptively subtitled *Books That Encourage Personal Independence in Harmony with Nature and the Environment.* A number of older Rodale titles can still be obtained through the Storey catalog.

I can't account for the timing, but I do know that the germ of this desire had been planted by lucky happenstance. As a student at Hanover College in southern Indiana in the early Sixties, I had met Anna and Harlan Hubbard, who had then been homesteading in Payne Hollow on the Kentucky shore of the Ohio River for about ten years. At that time, Harlan, who was in his early sixties, had published one book—*Shantyboat*—and had been painting all his life, but he was not yet widely known as a painter, writer, or "experimenter in living." The first time a teacher of mine brought me and my roommate the five miles downriver to meet the Hubbards, I hardly knew what I was seeing; but over many subsequent visits and years of friendship the spirit of Payne Hollow, and the lives being lived there, sank into me as deeply as anything ever has in my life; and that fall of 1984, for no reason very clear to me, the seed decided to sprout.

My timing couldn't have been much worse. We were committed to leave the following summer for England, where my husband, Ted Irving, was to head up the English Department's Junior Year Abroad program at King's College, University of London. Ted was not averse to the idea of homesteading even though it would mean moving farther out from the city—we were both accustomed to the long commute by train—but there was no way to pursue the idea until the year in England had been completed. So, reluctantly, I packed all my recently acquired Rodale gardening and homesteading books into boxes and shipped them to England, where they became the basis of my first science-fiction novel, *Pennterra*.

Immediately after returning to this country, in July 1986—even before we had moved back into our house—we spent a couple of weeks driving around the countryside within a fifty-mile radius of West Philadelphia, looking for "the homestead": fifteen to twenty acres of country land situated (a) close enough to Philadelphia for the commute to make some kind of sense, (b) far enough from Philadelphia that backhoes wouldn't be poised at the verges of cellar excavations everywhere you looked.

It was an idea that had come too late. Even to me, this truth became plainer and plainer as our search took us farther and farther north and west of the city. Really, it had been plain enough on the very first day. We had visited the Rodale Research Center at Maxatawny, near Kutztown, a number of times, and both of us loved the Pocono Mountain piedmont country where the center is located. So we began by driving north of Philadelphia, on the back roads of Bucks and Berks counties, heading north through the town of Oley—Wallace Stevens country—toward Kutztown. Ted was driving; I kept watching out the window for the end of development and the beginning of a settled rural landscape. Yet no matter how far behind we left the city, we could not seem to put the backhoes and raw earth behind us. In 1986 the recession had not arrived, and housing starts were still springing up like mushrooms—all the way between Philadelphia and the Kutztown-Allentown axis. All along that way, the evidence of farmland sold to developers was everywhere. There *was* no "settled rural landscape"; there was instead a countryside in the grip of a change of character.

I had not expected this at all and was very downcast. We cheered up temporarily when we found a wonderful house between Kutztown and the Rodale establishment, a house part Colonial stonework and part modern, which came with twenty acres of wooded hills on both sides of the road. A stream ran right through the property, and it had a fenced paddock and a little barn. And while the price was higher than we had intended to pay, it was less than what we ultimately ended up paying for our present home in the Philadelphia suburbs.

Two considerations prevented us from buying that house. The first was nervousness about the future of the area; the house, like most Colonial houses, was built right next to the road, and the light traffic of 1986 seemed certain to increase dramatically in the immediate future if development in the area proceeded as it seemed certain to do. The other was the realization that with the Reading-Philadelphia train line no longer operating, the commute could involve taking the Greyhound bus, or else driving an hour to Lansdale to catch a train to Philadelphia for a ride of another forty-five minutes or so: too much, too long, too exhausting, inconvenient, expensive. At that time Ted, then sixty-three, planned to retire at sixty-eight. (He decided later to defer retirement till seventy.) Whatever we did, he or we would be doing it for at least five more years. Five years of commuting between Kutztown and Philadelphia was an insupportable prospect. We had to accept it: Kutztown was out.

Phase Two of the homestead search took us into Lancaster County. Amtrak trains running between Philadelphia and Lancaster stop at a couple of other towns; this struck us both as a more workable—if still costly—alternative. The best farmland, of course, was too good for us; and most of the prettiest land is owned by the Amish, who would be wonderful neighbors for organic homesteaders except that those farms are not among the ones being subdivided and sold off in lots the likes of Ted and me could afford.

We connected with a Lancaster realtor, a former Amishman named Stolzfus, who had worked for one day as an extra during the barn-raising scene in *Witness* (real Amish people are prevented by their religious beliefs from being in movies, so they used people like our Mr. Stolzfus, who spoke Pennsylvania Dutch and could use carpenters' tools). We took a room and watched *Witness* on the motel TV. The next morning Mr. Stolzfus drove us around to look at several parcels of land, farms that were being split up.

The particular piece offered for sale sometimes had a field and outbuildings but no house, sometimes a house, barns, assorted other structures crammed from end to end of the acreage for sale. An enormous set of power lines from the Peach Bottom Nuclear Facility on the Susquehanna loomed above one farm-fragment. On another, pigs were being raised in darkness, shoulder deep in filth. Farms were being broken up at a terrific rate, but there seemed to be no attempt on the part of local governments to protect the character of the land by zoning; mobile-home parks and auto body shops were scattered about haphazardly among the "farmettes," and the whole area showed great instability.

It was all desperately exhausting and depressing. After three days we had to face the fact that we were not going to succeed at finding an affordable place we would want west of Philadelphia, either.

Helped by a local realtor, we tried Chadds Ford—Andrew Wyeth country—and Glen Mills. Everything, literally everything we were shown, was either hopelessly out of our price range, threatened by development, or both.

It was at this low point that our Five-Year Plan was conceived.

The Five-Year Plan required us to acknowledge that my dream of finding a small homestead within reach of West Philadelphia had come twenty years too late. We would confront this bravely, buy a house with a nice big sunny flat yard on the Media train line, garden up a storm, and bide our time. In five years' time, when Ted retired and we no longer needed to stay linked to the University of Pennsylvania, we would sell this house and find something, somewhere, closer to the idea of a homestead to buy. I was already teaching part-time at Penn; I could easily teach part-time someplace else.

With this in mind, I spent the fall of 1986 selling our old house and going around with our realtor to look at other houses in the same area. The one we chose was almost perfectly suited to the Five-Year Plan: a beat-up ranch with a wonderful—if desperately overgrown— backyard and a very prestigious address. From its leaky roof to its cat-urine-saturated carpets and floors, the house was in truly terrible shape—the only reason we could even consider buying it. But if "location, location, location" had any validity, we couldn't fail (we thought) to do extremely well in five years when we sold it to buy our homestead. Charged up with ideas about how the house could be renovated, and full of confidence that we could do a lot of the work ourselves, we closed the deal.

(Life has a way of upsetting plans of this kind. The first upset, a medical crisis, came in April 1988, when I was diagnosed with breast cancer and underwent a mastectomy and six months of chemotherapy. The second upset, the collapse of the real-estate market later that same year, was minor compared to the first, but still it delivered a major blow to the Five-Year Plan.)

In December 1986, ignorant of these pending developments, we moved in. The following spring we hired a guy with a tiller to till up part of the back lawn, and I built raised beds and planted the first big garden of my life.

I remember writing to friends somewhat later that summer that I hadn't quite taken in the difficulties inherent in simultaneously try-ing to (1) bring a very badly neglected eight-room house back into shape, (2) make and manage a large garden in a new location for the first time, (3) keep up a steady output of science-fiction stories, and (4) prepare to go back to a full semester-load of teaching right after Labor Day. Unwilling (renovating, gardening, writing) or unable (teaching) to give up any of the four, I struggled along in a very frustrating way. That season's garden was remarkably successful, but it was *big*, and I was stressed out and overextended—so much so that more often than not I would have to *force* myself out to do a given task—picking beans, wa-tering—until further delay would have meant that the whole summer's hard-bought time and energy would have been invested for nothing.

This was not the spirit in which I had planned to do my gar-dening, and it distressed me constantly that I seemed unable to stay on top of things better.

The following, or chemotherapy, summer was also the record-breakingly hot summer of 1988. The great heat made the debilities of chemo even more difficult to transcend. Between the biopsy on April 15 and the mastectomy on April 25 I had managed to get a lot of plant-ing done, and long before I could straighten out my left arm again I was spading up the raised beds and transplanting cabbages and tomatoes. Even that year the garden produced a surprising amount of food, considering the circumstances; but I meant to do better the following year, and I did.

Yet in both 1989 and 1990 the same problems—too much to do, too much garden to manage properly in the time available—continued to plague me. I loved the garden, but it was far from being an unalloyed joy; much too much of the time I spent digging and weeding in it was spent in the spirit of get-done-and-get-back-to-work-on-the-book. It was no longer possible to pretend that things would change with the passage of time; all too plainly, I was gardening either too little or too much. I needed either to find a way to give that part of my life more space, so time passed in the garden could be passed in the proper spirit, or I should scale back in order to achieve the same goal. That summer of 1990, I was working very hard on my third novel. Scaling the garden back seemed indicated; yet my reluctance to accept this conclusion was so strong that I stuck fast between the rock of passionate desire and the hard place of reality. And I kept on gardening badly and feeling guilty.

Then, sitting on the patio one July afternoon with a cup of tea and a new issue of *National Gardening*, I was struck by a brainstorm: what if I were to *clear out* a whole year, do no writing or teaching at all, so as to give *all* my energies to the garden? What if the garden and the one-acre lot it sat upon were to become—at least for that one year—the "homestead" of my heart's desire?

The thought excited me profoundly, but I was not so dazzled as to be purblind to the financial problems it would create. Ted and I, who were sixty and forty when we married and have no children, have never pooled our money; each month we do the accounts and split the expenses down the middle. Each month I was going to have to come up with my half of these expenses, unless we could work out a deal.

Ted was sympathetic, and the deal was struck. In exchange for what the more intensively managed "homestead" would contribute to the household economy by putting food on the table, I was to be let off paying my half of the mortgage throughout the homestead year, an arrangement which would both save my pride and conserve my resources.

I completed my novel-in-progress and moved the science-fiction files to the back of my file cabinet; I notified the university that I would not be teaching my usual courses in the fall '92 semester. These were good gestures, psychologically helpful, though in both areas significant compromises rather quickly proved necessary.

I met with two independent-study students once a week all through the spring, and with another during the second summer session; and each Tuesday during the fall semester, I taught a creative-writing course in short science fiction. The inducement was money, of which I'll have a great deal more to say in due course. I held the teaching to one day a week instead of three—an important difference—but best of all would have been no teaching, period.

Besides teaching, I had a career in science fiction to maintain in what might be called viable stasis, till I should be ready to take it up again in an active way. Specifically, it was necessary to accept several guest-author invitations, attend the occasional convention, keep up with editors and other writers, take an active part in publicizing my new novel, *Time, Like an Ever-Rolling Stream,* when it appeared from St. Martin's Press in September. I had hoped to do none of these things, or almost none, but the penalty for avoiding them would have been higher than it made any sense to pay—very much like digging, fertilizing, planting, weeding, watering, staking—and then failing to pick the beans till they're too stringy and tough to enjoy. I did accept this reality, though with a lot of griping.

Nevertheless, I would advise anyone interested in living out such a dream as this to put some savings by beforehand. Worrying about money, and accepting compromises in order to make some, won't ruin the homesteading experience—real farmers do this constantly—but I'm very sure it's better in every way not to have to.

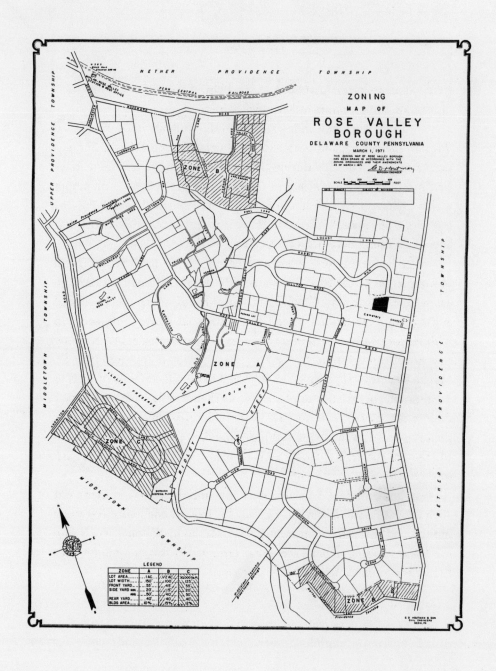

ZONING
MAP OF
ROSE VALLEY BOROUGH
DELAWARE COUNTY PENNSYLVANIA

MARCH 1, 1971

THIS ZONING MAP OF ROSE VALLEY BOROUGH
HAS BEEN DRAWN IN ACCORDANCE WITH THE
ZONING ORDINANCES AND THEIR AMENDMENTS
AS OF MARCH 1, 1971.

BOROUGH ENGINEER

SCALE 0 200 400 600 FEET

LEGEND			
ZONE	A	B	C
LOT AREA	1 AC.	1/2 AC.	30,000 sq. ft.
LOT WIDTH	150'	100'	125'
FRONT YARD	55'	45'	55'
SIDE YARD MIN.	20'	15'	20'
ONE +	40'	40'	50'
REAR YARD	40'	40'	40'
BLDG AREA	10%	15%	12%

G. D. HOUTMAN & SON
CIVIL ENGINEERS
MEDIA, PA.

February

February 14. The Homestead Year achieves liftoff!
The day began inauspiciously enough, with a glum
morning spent working on taxes, but improved steadily.
I met my friend Liz Ball for lunch; then Liz came on
back to the house while I went to buy a valentine for
Ted. When I arrived home, Ted and Liz were standing
in the short, dry, yellow-brown grass of our backyard,
contemplating the overgrown landscape feature we
refer to as The Island.

      Some long-ago owner of this house probably
spent a lot of money having The Island created by professional landscapers, the basic design is so attractive
and well thought out: a grouping of ornamental shrubs
—forsythia, azaleas of several kinds, a large holly tree,
and something that makes hard red berries that hang around all winter—all this arranged in descending order of size when viewed from
the patio, from the ten-foot holly down to the border of boxwood.

      At ten feet by twenty by thirty, The Island has swelled and
spread. It's much bigger now than it used to be even five years ago, when
we moved here. Its original lines have also become less and less dis-
cernible, as volunteer plants shoulder in and make themselves welcome
among those that were put there on purpose. Wild cherry trees, pokeweeds, and blackberry canes "planted" by birds have sprung up and
flourished; and the whole thing, nursery plants and wild ones alike, has
been laced democratically together by tough honeysuckle vines whose
foliage makes a rumpled green coverlet over the entire planting,
summer and winter.

      For five years, every time I've looked out at The Island, I've felt
guilty. Somebody lavished so much care on it once. I should get out
there and grub the honeysuckle out, prune out the deadwood, restore

things to their former glory. I've never done it. This whole one-acre property is The Island writ large, a jungle of humongous shrubs, Japanese holly and ordinary holly and rhododendrons and azaleas, beautiful, robustly unkempt, overwhelming. In 1986, when it was put on the market, the house had nearly been swallowed alive by its own thirty-year-old foundation plantings. The former owner, acting on advice from his realtor (which he had not entirely comprehended), had taken a chain saw and brutalized all the azaleas, rhododendrons, holly and forsythia into a yards-deep shelf bristling up the walls of the house to a point just below the windowsills.

Faced with the task of restoring all of this to acceptable size and shape, we began digging things up by the roots instead and starting over. But The Island, far from the house, was allowed to lead its own life of dishevelment and neglect.

Anyway, the place has been a haven for birds and rabbits, of which the yard always supports a great many.

Now that the Homestead Year has finally arrived, however, I expect to have more time for outdoor projects than ever before. One of the outdoor projects I've got in mind is the digging of a fishpond on which I want to raise a few ducks. After considering the needs of water plants, and the possible complications of duck raising in suburbia— even here in Rose Valley, where every house has an acre of land and plenty of privacy and elbow room—I concluded that the best place to put the pond would be behind the garden, as far as possible from the neighbors on either side and clear of both major oak trees, in a site that receives full sun at least six hours a day in summer. The above precisely describes the location of The Island, and so it was decided: rather than blunt-cut either of the splendid big oaks to provide more sun, I would take down the impenetrable tangle of brambles, vines, and tough twisted stems and dig the pond where all of this had been.

This is what Ted had been explaining to Liz when I came up to them. Both turned to look at me. "Liz says she and Jeff have a reciprocating brush-cutter that would go through this whole mess in twenty minutes," Ted said. I stared at him, taken aback. "But I told her you probably wouldn't want to do that," he went on. "I mean, after all, it's a *power* tool." He rolled his eyes at Liz.

Liz, who has her own pond, complete with waterfall and spawning goldfish, said, "You realize this is the worst place in the whole yard for a fishpond, don't you? What's wrong with just digging one out here in the grass?"

These skeptics were my husband (whose valentine I was clutching even as he rolled his eyes at Liz) and my good friend, an experienced garden writer and photographer. If *they* thought I was crazy, what are my chances of being understood by those who don't know me so well or feel so much sympathy with the impulse that led me to undertake my project?

The point of this year I've cleared out for homesteading (as I reminded them both) is *not* to do a certain amount of work as quickly and efficiently as possible, using whatever means might be made available. The point, corny but true, is to find out how much I can accomplish *without* spending a lot of money and without burning any more fossil fuels in the process than absolutely necessary. In other words, I *want* to take down The Island using the pruning saw and hedge clippers because I want to find out what it's like not to fall back on some machine the minute you face a tough job. Like the Old Order Amish farmers who farm as they do in order to maximize their closeness with nature, I can best accomplish my goal by behaving as though these machines did not exist.

My goal for this particular afternoon was to boil a roasting pan full of sap from our Norway maple tree—about three gallons—down into maple syrup. The tree is a couple of feet thick, big enough for three "taps," and that's how many I'm running. You put in a tap by drilling a largish upward-angled hole through the bark, into the soft wood of the tree—I use a half-inch bit on a cordless drill—and tapping a hollow metal spile into the hole. The spile has a hook underneath to hang a bucket from. You clamp a special lid over both bucket and spile to keep out rain and melting snow. After going inside to change, I removed the lids one by one, unhooked the three buckets from their spiles, and carried them to the fire circle near The Island. Last night was pretty cold; the sap inside, which is mostly water, was still frozen solid.

On Sunday I'd boiled down the first batch of the season, and half a garden-cartful of leftover wood was still there in a pile, ready to be burned. The roasting pan had not been cleaned from the earlier boil and was still crusted with soot on bottom and sides, but it was already 1:30, and I knew I'd be lucky to finish by dark. So, after getting the fire going, I carefully dumped two of the big rounds of frozen sap into the crusty pan and just set it back over the flames, edges resting on the cinder blocks I'd positioned there on Sunday.

Once the fire was well underway and the sap beginning to melt in the pan, action slowed to a crawl and life attained, for this little while, a beautiful simplicity. During the next four-plus hours I had virtually nothing to do except keep the fire well fed on the trash wood we save for the purpose—in this case, cut lengths of old trim from Ted's recently remodeled study, varnished and therefore damaging to the catalytic unit in the woodstove if burned there—and circle round and round the pan and fire to escape the smoke. As the white niter—saltpeter—in the sap foamed to the surface, I would skim it off with a slotted spoon. I'd discovered on Sunday that the sap would stay at a rolling boil if I held the fire's heat in by propping shingle-shaped pieces of wood on either side of the cinder-block "grill" so the wind couldn't blow the flames from beneath the pan. But these two extra things to be aware of and deal with didn't make the task very much more complicated.

Ted went back to grading papers. Liz shot some pictures and went home. I changed my coat for a light jacket and stood in the warming day mesmerized, first, by smoke and flames and then, gradually, by steam, as the sap heated and the big round ice "cubes" melted down. When there was room, I stirred from my reverie and added the third round hunk of frozen sap to the pan.

Time passed. For much of it, my mind was a peaceful blank— a state my "normal" life hardly ever lets me achieve. For a long, long time, the level of the boiling sap never seemed to lessen appreciably.

For some of that time I meditated on the ethics of burning up so much wood, producing so much CO_2, for the sake of less than a pint of mediocre syrup; but actually my mind is pretty straight on that subject. Ted and I take pains to avoid burning gasoline—we use public transportation, walk the mile to and from the local commuter train station in all but the vilest weathers—and we scrub the smoke produced by the woodstove. Only the central part of our house is heated by the oil-burning furnace. But the CO_2 locked up in wood—a renewable resource—is destined, unlike that in oil and gas, to wind up in the atmosphere sooner or later in any case. So, in exchange for conserving gasoline and fuel oil, I occasionally give myself permission to burn some wood.

For some of the time I thought about Liz and Ted, and their siding against me—in however friendly and teasing a manner—on the matter of doing without any tool that requires a source of power. I know what a conundrum I am, a science-fiction writer temperamentally at odds with the modern world. I bought my first car when I was nearly thirty. I was the last writer I know to acquire a computer, and we still don't own a microwave. I don't *like* machines. Nobody else I spend my time with appears to feel like this or understand how such a thing could be. Noel Perrin, who gave me my three sap-bucket lids and the spiles I'm using now—a writer after my own heart—*he* understands it, but he's in Thetford Center, Vermont, and I hardly ever see him. Wendell Berry understands and shares my deep mistrust of a society that doesn't know how to accomplish a damn thing without a machine to help it; but Wendell Berry is faithful to a lot of other values I don't share.

What snapped me out of the reverie after a time was the realization that my bees were also on the boil. On this first mild sunny day in several weeks, they were pouring joyously out of the hive. I'd been wanting a chance to open the hive for a quick midwinter inspection—all the bee books tell you to do that, although, this being my own first midwinter as a beekeeper, I'd never yet performed this chore—and the weather was the best it had been or might be for some time. So I stoked the fire well, then went inside and got my coverall, helmet and veil, smoker, hive tool, frame gripper, a bag of egg cartons torn up into little pieces, and brought them all outside.

After stuffing a wad of newspaper and then many pieces of egg carton into the smoker—a tin can with a lid shaped like a sideways funnel, attached to a small bellows—I suited up, tucked the skirt of my veil into the collar of the coverall, drew my gauntlets of pigskin and canvas over my sleeves, and walked back to the fire carrying the pieces of equipment. I hadn't handled this gear for three or four months and felt a little awkward; but it was no trick to get a fire started in the smoker and, as always, once the smoker was going, there was no reason not to tackle the hive.

I'm a little afraid of the bees. One day last summer, when I was still wearing some hand-me-down all-canvas gauntlets, a bee stung me on the knuckle of my right hand, right through the canvas. I was in the middle of inspecting the hive, with both covers off and a frame full of honey gripped in the gripper; for a couple of minutes I could do nothing about the sting. Also, I was curious as to how much reaction I would have to the venom; this is important information for people who want to keep bees. So, even after I had reassembled the hive and gone inside to inspect the damage, I didn't ice the sting or take antihistamines or do any of the things you can do to minimize the impact of being stung.

My hand swelled up like a pillow as far as my wrist, but there the swelling stopped. The doctor called it a mildly allergic reaction. I wondered whether my career as a beekeeper was going to be over before it had really begun, but he said not. He did advise me to buy a bee-sting kit, and to get his nurse to show me how to inject myself with a hypo of adrenaline in case of possible future need. I took the advice. But even thus prepared, and even though my reaction was not anywhere close to anaphylactic shock—and even though I ordered my leather gloves that very day—I've been apprehensive about working the bees ever since, and always feel a little trepidatious as I set about it.

My hive of bees is a swarm captured last April by Jim Castellan, my neighbor across the street, with my eager but most inexpert help, and is still housed in what is mostly his equipment (plus some of

Liz's and her ex-husband, Jeff's). Temperamentally, my colony appears to be on the irritable end of the spectrum; but the queen is a good layer, and since mine was a second swarm from the parent colony that spring, she definitely hatched out only last year. Usually queens are good for two years. I'll probably requeen that hive this fall, and meantime will put up with crabby bees.

Back when I got them I didn't know that I could (and should) have fed the bees on sugar syrup till they had "drawn out" all the comb foundation I gave them—that is, till they'd built honeycomb on both sides of the flat sheets of wax supplied by me. If I'd fed them, I'd have been able to take some honey off last year. As it is, I've yet to harvest so much as a drop, let alone a crop of honey, from the little beggars; but this summer, if all goes well, I should be able to take off enough honey to make this maple-syrup caper look as preposterous as it probably is.

I walked behind the hive, which is swaddled in black building paper for the winter, and puffed some smoke into the entrance where the bees were clustering and flying in the sunshine. After a minute I lifted off the outer cover and laid it upside down on the grass. There were a few bees on top of the inner cover, and more visible through the oval hole in its center. I puffed some more smoke through the hole in the cover, which the bees had glued to the hive body with the gummy stuff they make from resin, called propolis. Then I pried the inner cover off, and laid it aside too after checking to see that none of the bees clinging to the underside was the queen.

There were *lots* of bees in there. "Don't disturb the cluster," say the bee books of the midwinter inspection, but they weren't *in* a cluster; they were all over the tops of all the frames in the hive! The hive body resembles a hanging file cabinet containing ten files, called frames, suspended from the upper edges. When you take off the inner cover you see a box full of frames. To inspect one of the frames, it's necessary to pry it loose at either end—the bees propolize everything to everything—grip the top with a frame gripper, and *slowly* pull it straight up and out.

I wanted to know if there was enough honey left to see the

colony through the rest of the winter. On top of the frames was a mesh bag full of pieces of menthol to control tracheal mites, put there last fall and propolized tight to the middle frames. Tracheal mites, tiny parasites that infest the breathing tubes of bees, are a deadly menace, especially in winter; it's very common to lose an untreated hive to them (and sometimes a treated one).

The menthol balls had dwindled like mothballs, which they resemble. I didn't want to unstick the bag at this point in the season. So, using my hive tool—a piece of metal bent like a short, thin crowbar—and trying not to jostle everything, which isn't easy when all the parts are glued together, I gingerly loosened and inspected a couple of frames, to find out how the hive was fixed for food.

My bees did still appear to have some honey, but not a lot. The frames I pulled showed a similar pattern: an oval area of wax-capped honey in the center of the beautiful golden *empty* honeycomb. The perfection and beauty of the empty comb were very striking. Both sides of the frames showed this oval pattern of use; the bees obviously begin on the honey at the outer edges and work their way inward.

Bees were beginning to sting my gloves and clothes. I smoked them some more—*whuf whuf whuf* with the asthmatic smoker—but having learned that they were all right for honey at least for a couple of weeks, I decided to exit gracefully and check again toward the end of the month, on a day when no other project needed my continuing attention. Or that's what I told myself. I like keeping bees, but would prefer gentler ones. The books, and the magazine *Gleanings in Bee Culture*, all have numerous photographs in them of people handling bee-crowded frames barehanded. Obviously, a gentler strain of bee than mine exists. Something to plan for in the future. What you want is a gentle queen. Every bee in the hive is the queen's offspring; gentle queen, placid hive. Though they do say—it may be folklore—that meaner bees make more honey. Which would be consistent with the fantastic yields of the Africanized bees, the so-called "killer bees," now working their way north of the Mexican border.

I put back the frames and covers, collected my tools, and walked back to the house. Angry guard bees followed me, and half a dozen moribund guard bees, those who had given up their lives along with their stingers, clung to my sleeves and hatbrim. The flying bees gave up before I got back to the house. I brushed half a dozen others off at the door; they fell onto the patio, dying. Before pulling off my helmet and veil in one piece, I checked as best I could to be sure there were no more bees around. Once last summer I brought several into the house with me by mistake, an error none of us was happy about.

But the coast seemed clear, so I drew down the long zipper of the white coverall, climbed out of it, tossed it and the helmet onto the dining room table through the door, and returned to my fire.

This had burned down considerably, but the coals were very hot and the sap was soon bubbling merrily again above the flames. And now at last the white niter scum on the sides showed how the level had dropped, and I could see the first faint hint of yellow in the rolling liquid. It was 4:00 P.M. I took up my vigil again, stoking the fire, keeping shingles propped on either open side to conserve heat, automatically circling to escape the smoke.

The Island was so close to the fire that, after another mindless interval, it struck me that I might as well get started hogging brush.

Ted had already cut down the volunteer wild cherry trees for woodstove fuel, and the big holly in back; it's a point of pride with him to harvest all his stovewood from our own tiny woodlot. The remaining plants were smaller in diameter, but impenetrably dense. After sizing up the situation I fetched out the hedge clippers and a pruning saw from the garage, then went back to get some leather gloves—a lot of the outside layer of The Island consists of wild blackberry canes, which you can't handle with anything less. As with bees, so with brambles.

Eyeing the tangle without spotting a logical place to begin, I finally just grabbed the hedge clippers in two hands and started chopping, straight in and fairly low. Big masses of branches and vines sagged away from parent stems; I dropped the clippers and hauled these away, chopped again, hauled again; before long I'd built quite a respectable pile of brush and there was a gaping hole in the integrity of The Island.

I began systematically cutting and pulling honeysuckle vines off the other stuff. It was necessary to keep stopping to stoke the fire and check the color and level of the sap, which now was beginning to yellow up and drop in earnest (and to leave a definite sheen of sweetness on the slotted spoon). Nevertheless, by the time the liquid in the roasting pan was down to half an inch, and I had to quit for good, it had become exhilaratingly obvious that the task of cutting down the whole Island by hand would by no means be beyond my ability to do all by myself.

But the syrup was ready to be moved inside, so I had to put down my tools, shuck my gloves, and dash back to the house for a saucepan. Setting it on the ground, I used the gloves to pick up the roasting pan by the handles and pour in the dirty yellow liquid, full of sterilized bark bits and black flecks of fly ash from the fire. Empty, the roasting pan went upside down, one end propped on a rock, and the saucepan with its pool of golden treasure went inside to the stove.

It was 5:30 and beginning to be twilight. Feeling wonderful from my afternoon outdoors, I stood at the stove in boot socks and muddy-kneed jeans, cheeks tingling with windburn, and floated the candy thermometer in the boiling almost-syrup. You're supposed to check the boiling point of water on the day you boil—which, depending on weather factors, might or might not be exactly 212°F—then add seven degrees to that, and whatever you get is the temperature you have to bring the syrup to in order for it to be really syrup. But in practice you don't actually have to test boiling *water* because the syrup boils at the same temperature as water until it's almost finished—this day, at least according to this thermometer I don't entirely trust, at 216°, meaning we would have to take this batch all the way to 224°.

Outside it grew steadily darker; through the kitchen window I could see the coals in the fire pit, glowing beautifully. Ted offered to go out and douse them while I tended to the syrup making, and I agreed, though it felt like a crime to waste such hot and beautiful coals; it seemed there should be something we could roast for dinner over or under them—turkey franks, potatoes—but there wasn't. The syrup was now boiling in such a way that large bubbles covered the entire surface, making it look like a round yellow disk of bubble wrap; at this stage it can boil over easily and must be monitored closely. The thermometer, held a bit off the bottom of the pan, in case the metal in contact with the burner should be hotter than the liquid, was finally beginning to climb.

Finally at about 6:30, when I squinted through the steam-coated thermometer casing, my best guess was that the mercury did read something like 224°. The syrup seemed slightly thicker than the first batch. I suddenly panicked to think I might have overboiled it; this happened the first year I tapped my trees, and resulted in finished syrup whose texture was somehow *brittle*—I'm not sure now why "brittle" seemed to apply to a substance so viscous—and which, soon after, began to precipitate into rock-hard crystals glued to the bottom of the jar. Not wanting this to happen again, I snatched the pan off the stove.

Noel Perrin had given me a genuine felt maple-syrup filter. I was with him when he bought it at a real Vermont hardware store, so I know for a fact that it is the genuine article: a blunt cone made of thickly felted wool, fifteen inches in diameter and fifteen deep (after washing), with four tabs to hang it from sewn to the cardinal points of the rim.

I hadn't filtered the earlier batch of syrup through Ned's classy filter because I hadn't read the directions in time to realize you were supposed to wash the thing before the first use. Now I had a clean filter, but was faced with two other problems. The first: the directions that came with it said you should soak the cone in boiling sap before pouring in the syrup.

So far as I knew, all the boiling sap within miles had been transformed to unfiltered syrup on my own kitchen stove. I pondered the probable reason for soaking the felt cone in what is for all intents and purposes boiling water and decided this instruction must have to do with the permeability of dry versus wet wool. In that case, I reasoned, hot water should work just as well; so I soaked the bottom three inches of the filter in that.

I say "the bottom" because that was my second problem: I had maybe a cup, maybe a little more, of syrup, and a fifteen-inch cone to pour it into.

Having turned the top foot of filter back like a gigantic cuff, and positioned the blunt wet tip of it, like the toe of a sock, into a funnel stuck into an empty honey jar, I poured the syrup—by now no longer really boiling hot—into the filter, where it formed a golden pool containing many black particles. But instead of draining straight through, like my previous batches strained through synthetic felt, the syrup filled up the filter tip to the top of the funnel and simply sat there. Then, ever so slowly, it began to drip into the honey jar, one golden drop at a time.

I tried stirring the syrup to make it move through more briskly, but to no avail. I tried heating the remaining syrup in the pan, in case it had cooled too much for proper filtering, but that made no difference. In desperation, I switched to my old faux-felt standby, but to my surprise the stuff didn't want to go through that either.

Eventually the syrup did drip through. By 7:00, ten ounces of syrup, a bit more than a cup, in color the light pure gold of clover honey was in the jar, the black particles were in the filter, and I sat down to supper across from Ted, happy and tired and deeply contented with my day.

Ted's been casting peculiar glances at me all day. His comments to Liz about The Island probably reflects a certain apprehension: how much is this latest madcap adventure of mine going to interfere with his own plans for the summer? But I know—he said so—that he's also been remembering the first winter in this house, and our first maple-syrup adventure. There are slides of him standing out in the snowy yard where the garden was to be planted, coffee mug in hand, expression of reverie on face, minding the fire. Most of the fun for us both, that year, was in discovering that you could actually boil maple sap down into syrup just like they said—even if your maple trees weren't sugar maples, and even if it meant spending a lot of time and work (and wood) for a single pancake supper, however thrilling.

THE PROTO-HOMESTEAD
FEBRUARY 1992

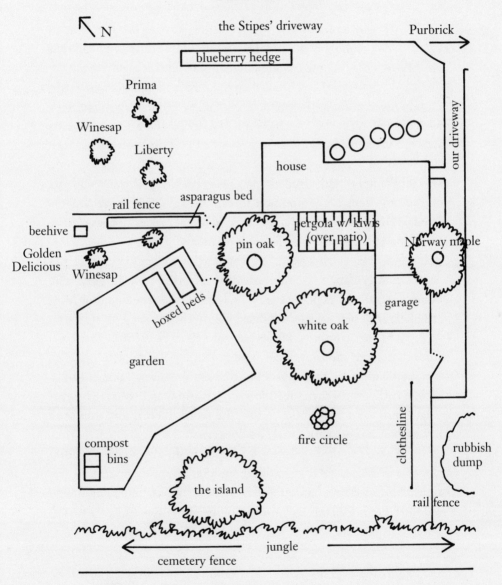

Making syrup is commonplace now, and Ted's too busy to work all day just to produce ten ounces of the stuff. Too bad. Doing it together was fun.

My own emphasis at present being not on product but on process, I count the day well spent. The problem so far has been that days like this don't come often enough. I cleared this year in order to have many, many, many days in which the simple straightforward labor of food production was to dominate my life, my thoughts, my energies and activities. Yet so far, 1991 has refused to let go of me. Commitments made last year, to do readings or visit classes at other institutions, have to be honored. Even though my actual teaching here amounts to only two independent-study students in creative writing, former students continue to require my attention in various ways. The novel I completed in October is in production now, and that process makes claims. From time to time, also, I go through periods of poorish sleep.

Another consideration is that into my Homestead Year has been introduced the two-step ordeal of reconstructive surgery. The January operation lies behind me now, but the one scheduled for May, at the height of planting, looms dismayingly ahead.

This list barely scratches the surface of the clutter that continues to prevent me from focusing on the Year. I've done a lot of information gathering, true; but all of that reading, note taking, library visiting, calling or writing for more information, and so on, has been wedged into the cracks of the "other" life that stubbornly continues to hold sway as my primary focus and concern.

There's relatively little I seem able to do about changing that balance and focus. It's my abiding hope that as spring comes on, the balance will shift in a natural way from desk, books, and computer to outdoors: digging, planting, building, moving and spreading straw.

March

March 20. This afternoon, while the sleet melted off the grass and the weather warmed, I celebrated the first day of spring by starting, let's see, fifty-seven seedlings—planting two seeds per pot if it was new seed ordered this spring, three if it was leftover seed that had spent the past two years in our freezer. Not every seed you plant comes up, and viability declines with age (though freezer storage helps), but that ought to get me at least one healthy seedling in each little pot. The crops: cabbages (20), tomatoes (18), sweet peppers (6), eggplants (5), and lettuce (8).

In principle, I'd rather not coddle seedlings under lights; I'd prefer the simplicity and unfussiness of direct-seeding everything. In principle, I would prefer to grow only open-pollinated varieties, too, and save the seed. In practice, I've always departed from both these preferences.

For one thing, tomatoes, eggplants, peppers, and brassicas take almost the entire growing season in this southeastern Pennsylvania garden to complete their life cycles if sown outdoors. This works okay for fall broccoli and Brussels sprouts, but I want tomatoes in July, not September. For another thing, our soil contains a lot of soilborne diseases; ordinary tomato seedlings, bought in flats from the local nursery and planted out in the garden, invariably die. To make a crop, we have to grow tomatoes with multiple-disease resistance, varieties with VFNT after their names in the seed catalogs. These multiple-disease-resistant varieties are all hybrids, and hybrids don't breed true from seed. I can't save tomato seed over from year to year unless I'm willing to be satisfied with whatever grows. We do commonly get numerous volunteers in the previous year's tomato beds, and from compost full of viable seeds from the previous year's crop; and some of these, if not weeded out, mature and produce fruit. But size and degree of flavorfulness of these volunteers are always a surprise, and usually, though not always, inferior. Sweet Chelsea, the toughest tomato variety I ever met, is particularly prone to sending up volunteers similar but not identical to itself.

The local nurseries offer seedlings of only a couple of the more popular varieties like Better Boy, not these tough VFNT tomatoes; nor do they sell seedlings of the savoy varieties of cabbages that make good coleslaw, and coleslaw is the reason I'm growing cabbages.

So there are some seeds I need to start every year. But I don't germinate the seeds in flats and pot them up when they have four true leaves, as many people do. This eliminates one step in the process.

Lettuce *can* be direct-seeded; but, in addition to soilborne diseases fatal to members of the nightshade and cucurbit families, we also have a huge number of slugs, which eat the young lettuces faster than they can grow. As an organic gardener I can't zap these pests with poison bait, the most effective control; and beer traps, the most frequently recommended organic slug fix, have never worked well for me. (Slugs are supposed to drown in the beer, but mine prefer tender greens to Coors.) To get lettuce I have to set out husky adolescent transplants, that will grow faster than the slugs can eat them. When the ducks come it'll be fun to see how good a slug patrol they make.

In principle, I would prefer not to buy any seed-starting supplies; I'd like to mix up my own seed-starting mix at the end of the season and sterilize and save it to make soil blocks the following spring. But I'm starting out the Homestead Year flat-footed, so to speak, making do with what's on hand and patching in temporary solutions while I work on permanent ones, so in fact I did buy some Jiffy peat pots and seed-starting mix—peat and vermiculite, my favorite stuff for the job, though it's the devil's own task to get it to absorb water. Also there's a twenty-pound bag of potting soil left over from two years ago. The potting soil, which absorbs water just fine, is supposedly too heavy for seed starting. The last time I did this—two years ago—I ran out of seed-starting mix, couldn't find more at the garden center, and had to fall back on this darker, lumpier stuff full of perlite, which always reminds me of Styrofoam; so this will be my chance to see whether the cabbages actually do grow better in one than in the other.

In previous years I've made good use of Jiffy pots, those little brown hockey pucks of solid pressed peat in nylon netting which expand when you put them in water. They're very easy to use, and they spare you the stage of trying to wet the dry shredded peat in the mix. I've had good luck with Jiffy pots for everything but lettuce; but they're awfully little, and the seedlings have a tendency to get top-heavy and rootbound if I happen to start the seeds a bit too early. So I'm not going that route this spring. For today's work I used some plastic pots and trays that have seen many years' service; if you're going to use things made out of plastic, I always feel, it's important to be able to use them over and over. You can't transplant the seedling pot and all, as you can with Jiffy pots and peat pots, but you can wash and reuse the whole "system," and I've never noticed that the transplants grown in these little pots are badly set back by transplant shock when moved out from the basement to the garden.

I also used some ordinary square peat pots ordered from one of the seed companies—this instead of buying more plastic pots and trays. I'm hoping to shift completely over to soil blocks after this year.

Books about seed starting always makes it sound so hard and complicated. I just fill up all the pots with planting mix, saturate the mix with water, poke a few holes in with a pencil point, drop in seeds, and use the pencil point to close the holes. You're always warned not to cover seeds by more than twice their depth in soil—surely more of a general recommendation than a firm rule: how do you measure twice the depth of a tiny tomato seed in lumpy mix? I just press a little bit of mix on top with the pencil. The seedlings always seem to do fine.

When all the seeds were planted, I bundled them up in or under used plastic bread bags and put them where they would get the bottom heat that hastens germination. The little green Burpee twelve-pack of tomatoes and sweet peppers and the white plastic pots of tomatoes and lettuces in their black tray, in separate bags, went on top of the hot-water heater in the basement. The five eggplants in peat pots, set in an aluminum tray that once held an Entenmann cake, are also on the hot-water heater. There's no room for the cabbages—twenty peat pots arranged on an old metal tray—on top of the heater, so they're sitting on the floor of my study, in a spot where the bare wood floor is warmed by the hot-air duct just beneath.

These variety choices reflect years of trial and error, of discovering what will grow well here, and what Ted and I both like to eat:

Cabbage: Salarite, a delicious type for slaw and much less space-greedy than the equally delicious Savoy King. Stokes, the only company I know of that offers Salarite in its catalog, has been coating the seeds in pink fungicide. Bothersome to dissolve it off, but I do; there will be no chemicals in this garden. Twenty plants.

Tomatoes: Hybrid Gurney Girl VFNT; Beefmaster VFNT; Viva Italia VFS. The Beefmaster seed lived for the past two years in a glass jar in the freezer. It's been started in half the green plastic Burpee twelve-pack. The Viva Italia, a gift from a friend, is in four plastic pots which sit in sockets in a black plastic tray, the Gurney Girl in eight others.

Eggplants: Burpee Hybrid, a seed-order freebie, leftover seed saved in the freezer from 1990. Five plants in peat pots.

Sweet peppers: Jupiter. Said to be good. Six plants, filling the other half of the Burpee twelve-pack.

Hot peppers: Hungarian wax. I opened the packet (from Stokes), found the seed coated with the same pink fungicide as the cabbage seed, and threw the packet away. Lots of other sources for Hungarian wax peppers. After this year, apart from the Salarite order every two or three years, Stokes doesn't get any business from me until they offer untreated seed again.

Lettuce: Green Ice. Eight plants in plastic pots. Unfailingly successful in my garden. I'll start some Red Sails and Black-Seeded Simpson, too, when I get some more supplies.

March 23. The biggest snowfall of the winter fell yesterday; it's a soggy white world out there. The five apple trees are standing in snow, their buds tiny and tight. Inside in a bucket of water on the hearth, slender boughs from the two Winesaps have broken dormancy and begun to open their leaves. The Winesaps didn't bloom at all last year. This year, when I pruned the apple trees—a long process spread over two months, from the end of December to the end of February—I left the tall upright shoots at the tops of the Winesaps; then, in early March, I cut those shoots and brought them inside to "force" the buds to open, hoping to see flowers (finally).

Last winter, very hard at work on my current novel, and on publicity for the previous one, I kept putting off pruning the trees. When I finally got around to the job it was already early April. The trees were then four years old; dwarf apples often bloom at three, but the year before, 1990, only the Prima had produced so much as a single blossom, let alone matured an apple (one). I dealt with four trees as usual but decided to leave one—the Prima—unpruned, to see if by taking off too much wood I was preventing fruiting.

That spring of 1991 the Golden Delicious and the Liberty had their first few flowers, but the Prima was *covered* with blossoms, many of them on the thin outer tips of branches that certainly would have been clipped off if I'd pruned the Prima too. Our first apple crop was harvested from a messy-looking, overgrown tree, weak branches dragged almost to the ground by heavy apples—not the way it's supposed to be.

Obviously, I'd been doing something wrong. Last August, at the Rodale Research Center's annual GardenFest!, I attended a workshop on pruning apple trees and described my situation.

April, I was told, is far too late to prune. Apple buds initially have the capacity to swing either way. Whether they become flowers or leaves depends on a number of factors, of which the amount of sunlight they receive is probably the most important (hence the preponderance of flowers on the outermost tips of the branches). By early April, three-

and-a-half months past the winter solstice, the buds' destiny has been determined. I'd waited till the die was already cast, then gone forth and systematically cut off all the flower buds.

I was also amazed to hear at GardenFest! that the branches of dwarf apple trees like these are only supposed to grow twelve to eighteen inches a year. Ours commonly grow four *feet*, and sometimes six or seven—the result, says my county extension agent, of putting fresh grass clippings around them all summer long, every time Ted mows the lawn. Without fertilizing them at all in any other way, we're overfertilizing. There's a ton of nitrogen in grass clippings, even from such a laissez-faire lawn as ours. This year I'll mulch with straw.

I just ran down and checked the hot-water heater. Less than three days after being planted, two lettuce seeds have sprouted—and so it begins again.

March 24. This morning five of the eight lettuce pots have seedlings, as does one of the Gurney Girl pots. A problem of having put two different crops in the same potting "system" now reveals itself: the lettuce should come out into the light; the tomatoes should stay covered and the bottom heat should continue for them at least another day, till more of them come up. No time to settle this before leaving for a day in town, so I took the whole tray out of its bag and set it on the floor by the south-facing patio doors, to spend the day in the sun.

When I checked on the cabbages, seven of the ten planted in potting soil had sprouted, compared to only three of the ten planted in Jiffy mix. Moreover, three potting-soil pots had *two* sprouts. So much for potting soil being too heavy for seed-starting, though just the difference in *number* of sprouts could be accounted for by how much bottom heat each seed received; the floor's hotter under some of them.

These cabbages have come out from under plastic for keeps, and go into the sun with the lettuces. My basement arrangement of fluorescent lights isn't ready yet for sprouted seedlings; setting it up will be tomorrow's number-one priority.

When I got home from town, the packet of black currant plants was waiting on the porch. I left it where it was, to keep cool and be dealt with tomorrow—my number-two priority. For years I've wanted to grow black currants, source of my all-time favorite jam, largely unobtainable in this country (though extremely popular in England), but was stopped by the purported link between white-pine blister rust and all the *Ribes*, the family of plants gooseberries and currants belong to. *Ribes* species were believed to be an alternate host for the rust virus, and for that reason weren't to be planted within 200 feet of a white pine tree.

When our suburb was developed, forty years ago, fast-growing white pines were planted along many of the property lines. We have some; so do our neighbors on both sides. Huge as they are, white pines are peculiarly fragile trees. Every winter, a wet heavy snowfall brings down enormous branches which drag power lines along with them, block roads, require the services of tree removers. When a little tornado ripped through Rose Valley, two years ago, only the white pines were damaged; the path of the tornado can still be traced by the large snapped-off trunks of white pines—including one of ours, the falling top of which crushed one of our Winesap trees (not fatally by some lucky chance) when it fell. Another of our pines blocks the late-afternoon sun from the garden. On the whole. I prefer gooseberries and currants to white pine trees; but consideration for neighbors of a different persuasion has kept me from planting any.

Recently though I've been reading in various places that the link had been somewhat misunderstood, and the various *Ribes* species given a bum rap. When working out my seed order this spring, I tried to find some of these imperfectly remembered articles to see whether growing some currants might not be possible after all. Failing to find any, I turned to *Rodale's All-New Encyclopedia of Organic Gardening* (Rodale Press, 1992). The encyclopedia rehearsed the usual information about white pine blister rust and *Ribes* species, then advised "grow resistant varieties." Among preferred species they listed "Consort," a kind of black currant.

Most of the nursery catalogs I consulted offered only red currants, but The Miller Nursery catalog listed a variety of black currant called "Prince Consort"—which was exciting, since if this was a relative of "Consort" it looked like I might be in business after all. I called the Miller company. The customer-service woman I spoke to said yes, all *Ribes* were alternate hosts for the rust. I read the encyclopedia to her; she put me on hold and consulted Mr. Miller himself, who assured her that "Prince Consort" was indeed rust-resistant. Resistant in this case must mean not a *host* for the virus; in any case he told her, and she told me, that it would be perfectly okay to plant "Prince Consort" in the vicinity of white pine trees. "I didn't even know that myself," said Customer Service. Hooray! Currant jam at last!

(When I double-checked with my local County Extension Service, the agent gave me the same piece of received wisdom; but when I told her that Mr. Miller said "Prince Consort" is okay, she immediately said she had tremendous respect for Mr. Miller, and if he said it was okay, it was okay.)

March 25. Tonight the cabbage and lettuce seedlings are out of their bags and under lights in the basement, and the three black currants are planted.

About those lights: a few years back, on advice gleaned from numerous magazine articles and books, I bought a shop-light fixture, several four-foot fluorescent tubes, some lightweight chain, and two hooks with screw-in threads. The hooks were screwed into rafters above the worktable in the basement and the shop light hung from them by the chains. After waterproofing the table with some slit-open plastic trash bags, I lined up my motley assortment of seedling pots on the table and lowered the lights till the tubes were suspended a couple of inches above the plants. Then during the following eight weeks, as the plants grew, I raised the fixture up the chain, day by day and link by link, always keeping the light just above the crowd of seedlings.

It worked fine. Some articles tell you to use one warm-light tube and one cool blue tube, but I didn't do that. Other authorities say it doesn't matter, and indeed it didn't seem to. Because this way of growing young plants relies on electricity, I like the old windowsill method better; but this method did seem to produce sturdier, more upright seedlings. We plan to put up a solar greenhouse in time to stretch the present growing season into the fall, so next spring the seedlings should have better growing conditions than either windowsills or electric lights can provide.

This morning I carried a bucket of warm soapy water and a sponge down to the basement and washed the accumulated cellar dust off the black plastic still covering the table. I also washed the shop light and the fluorescent desk lamp I use as a backup light, and plugged them in. Then I carried down the trays of seedlings and tucked them in place. After unscrewing all the other bulbs in the basement, so we can turn the fixture off at night and on in the morning by the upstairs switch, I went back up and left the little green leaflet faces yearning upward toward the light.

After lunch Ted and I walked around the yard trying to decide where to plant the three black currant shrubs that will probably live at this address longer than we will. We settled on an area parallel to the fence, near the other fruit plantings. Then Ted went back to preparing his Shakespeare class, and I went to get a shovel.

The useful Rodale book *Backyard Fruits and Berries* is silent on the subject of currants, and the instructions from Miller that came with the plants were sketchy, to say the least, so I fell back on what I had already learned from planting the apple trees and blueberries. After undoing the careful wrapping—"state of the art," they proudly call it on the big envelope the currants came in—I put the three plants, already pruned so the roots and tops were in balance (according to the instructions), in a bucket of water to soak while I dug the first hole. The plants had already broken dormancy and showed a quarter-inch of green leaf at the bud tips, making me doubt the assertion on the mailer that they would keep up to three weeks in the packing material, but I have to hope the folks at Miller know what they're talking about.

I dug a round hole about sixteen inches across and a bit deeper, four feet from the fence line—far enough that the lawn mower will be able to get through, at least for a few more years. As the digging progressed, I dumped the excavated sod and heavy clay soil on an old shower curtain to keep the soil from sifting into the grass and getting lost. When the hole seemed deep enough I got two empty buckets and filled one with water and another with absolutely beautiful 1991 compost, saved all winter in a big plastic garbage can with a good tight lid. I dumped some compost in the hole, mixed it with loose soil in the bottom, and poured in half a bucket of water. When it had settled, I set the first currant bush in a slurry of compost and clay having the consistency (and close to the color) of mortar. Finally, after backfilling the hole with more soil and tramping it down, I poured the remaining half-bucket of water into the yellow clay saucer in the center of which stood the new bit of edible landscaping, wearing the slightly dazed look that I knew from experience would wear off pretty soon.

The O.G. Encyclopedia recommends a spacing for black currants of eight feet in all directions; the Miller directions called for "two short steps apart"—four or five feet. One of the things you learn when you start reading extensively among gardening experts is that even experts disagree, and there are lots of ways to do things "right"—right enough to get a satisfactory crop, in any case. I compromised and dug the next hole about six feet from the first along the fence line, and the third six feet beyond that.

By the time I'd covered the pile of unused soil against the forecast rain and put my tools away, this currant-planting project had consumed two hours, and I was exhausted. It's exactly as I already knew: if I spend the morning paying bills and puttering around doing assorted indoor chores, then go shopping, *then* go outside to work on some gardening project, the spirit in which I work is simply wrong. I dash at the task the same way I dashed in and out of the camera store and the hardware store. Having arrived at the site with just the shovel and the three currant plants in a bucket of water, I get exasperated at my inefficiency in having to return to the garage again and again for things I hadn't had the sense to realize I would need until the need arose: the plastic groundsheet, the trowel, the two other buckets for carrying water and compost. I stab the shovel jerkily at the ground, have trouble establishing a rhythm, wear myself out trying to hurry.

All of which is why being a part-time gardener on the scale I'd been doing it was not working very well.

To garden on this scale, I need the garden to be the focus of my energies and attention—thus the Homestead Year. Part of the point of gardening is the food you grow, but the rest of the point is more nebulous, having to do with spiritual values and satisfactions. I've been producing a lot of food, but the happiness I once derived steadily from a much smaller garden flickers in and out of existence in this larger setting, a casualty of the ratio between available time and number of necessary things to be done other than gardening.

The Homestead Year was to be my solution: scale the gardening activities *up*, not down, but give them the time and attention they require.

Only so far, and very much to my surprise, I *still* haven't really been able to make good on the commitment.

March 29. On the living-room hearth, the bucket of Winesap wands is showing a few beautiful mint-green leaves, but so far no flowers. Out in the yard, the buds on all five apple trees are still closed tight against this cold, late spring. Today after lunch I opened the new bottle of dormant oil, consulted the labels, and measured nine tablespoons into my yellow plastic sprayer, which I then filled to the gallon-and-a-half mark with the spray attachment at the kitchen sink (to aid this process, which goes against nature, of mixing oil and water). Then I trudged out with it to spray the trees.

Like spring, I, too, am late; ideally, this should have been done a month ago. But the right combination of dry weather, calm winds, and being at home has persistently failed to occur. Today was milder than it's been, in the upper forties, but a bit on the windy side of ideal; but I'll be in town all day tomorrow, so it was carpe diem or risk another major delay.

Bushed after a morning of organizing my study, I had to force myself to the task. I'm not crazy about mixing sprays of any kind—we don't have a deep laundry sink to do it in and the sprayer leaks at several important jointures, usually onto my pants legs or into my shoes—but once outside I found the task rather more pleasant than not. So pleasant, indeed, that I came back twice to refill the sprayer tank and finished the job, though I'd managed to do it at all by promising myself I could quit the first time I ran out of juice. And now all five trees are sprayed. I can cross that chore off my list.

The first spring I did this, a single sprayerful of oil-and-water sufficed to paint all five little trees with scale-insect-smothering dormant oil—it's the trees that are actually "dormant," not the oil. Now one sprayerful will just about cover two of them, and the biggest tree, the Prima, takes an entire one-and-a-half gallons all for itself.

The tornado-damaged Winesap and the Golden Delicious tree are both close to the beehive, and the bees were doing some lively buzzing in the sun-warmed air in front of the hive. I was careful to keep the spray away from them, but several times bees whizzed past my head; it appears that deciding not to locate the fishpond between these two trees, and thus so close to the hive, was prudent. At a minimum we would have been in difficulties during the digging.

I also attempted to stay downwind of the spray myself, but occasionally caught a bit of fine mist in my face. Later I took off my glasses and found tiny specks of oil on the lenses, carried by the mist. Encouraging, to see that oil and water do mix, and all was as intended.

Later I'll have to spray again: wettable sulfur, which I've never used before, and/or liquid rotenone/pyrethrum from Gardens Alive! Disease has not been a problem for me in growing apples—all the trees except the Golden Delicious have some built-in disease resistance—but insects have been, and insect-damaged apples don't keep. It's difficult to grow apples organically. I don't object to some damaged fruit—the bad spots can be cut out and the rest turned into applesauce (or maybe cider, if I can locate a used press)—but I'd also like to carry some through the winter for fresh eating, and that'll take sound apples.

It's been a week now, and only one of the Gurney Girl tomatoes, and *none* of the Jupiter peppers, is up. Last night I faced this fact, sighed, plucked the seven unsprouted Gurney Girl pots from their sockets in the tray, and tucked them onto a plate under a plastic bag; and back onto the hot-water heater they went. This afternoon, one tiny tomato had sprouted! I put that pot back under lights and returned the others to their bottom heat. As for the peppers, foolishly put into the same seed-starter as the happily thriving Italian tomato seedlings, I'm not sure what I can do. Would bottom heat harm the tomatoes once they're up and growing? I don't know and hate to risk harming them; they look so jolly and frisky.

I've been worrying about the bees in this continuing cold and windy weather, and stopped across the street this afternoon to consult with my beekeeping neighbor, Jim Castellan, who told me he's begun feeding his four hives for the same reasons I was wondering if I should begin to feed mine. His fifth hive, a weak one, "starved out"—unnecessarily; he felt bad about it, knowing he could have saved them by feeding sooner. The bees are building up their numbers this time of year and need more honey (or a reasonable facsimile) just when overwintered stores may be giving out. I think I'll get ten pounds of sugar and start feeding this week. I'll check the hive first if it's warm enough, but if not I'll just cook up some syrup, and go ahead and give them a snootful on Tuesday. I would *really* hate to lose this hive.

It's hard to generalize about people who keep bees, except to say that they're a diverse and unusual lot. Jim, for instance, is a director in human resources systems at SmithKline Beecham, the pharmaceuticals company, and prior to that he was a member of the U.S. rowing team at the 1976 Olympics in Montreal. Before knowing him, I wouldn't have associated either of these jobs with keeping bees, but there you are.

Jim is the reason my beekeeping dream came true. I heard about his own long-standing fantasy of having some bees one day almost as soon as we moved in across the street from Jim, his wife Lynn Kelley, and their son Liam, then four. Not long after, he enrolled in a beekeeping course for some hands-on experience. Later, when a chance to acquire a hive from a neighbor came, he asked whether—in view of my own interest—I'd like to furnish the pickup for transferring the hive from the neighbor's yard to his own.

That was my first close encounter with bees—not a very close one; I had no equipment and observed the entire proceeding from inside the cab of my truck, along with Jim's little boy, Liam. Over the next couple of years I sometimes watched from a safe distance while Jim performed various hive manipulations, and I attended a couple of meetings of his beekeepers club. Last summer I even played eager assistant to his expert, when he got a call about an exposed hive—a colony of bees that, evidently having been unable to find a cavity to build their hive in, had hung it from a high limb of a big tree. A barefoot child had stepped on a bee and been stung. The family wanted the bees out of their yard and weren't willing to wait for winter, when any exposed hive at this latitude would freeze to death. To my mind, the rounded, parallel lobes of honeycomb resembled nothing so much as the fins of a hot-water radiator. The hive was covered with vigilant guard bees and was very beautiful.

Jim is a big, tall guy, and this was an awkward job. Crouching atop a long, long ladder in his coverall and veil—straddling a dead branch as a matter of fact—and juggling his tools as best he could, he cut the lobes of comb loose from the limb and caught them in a cardboard box. Back on the ground, we used string to tie the pieces of comb into frames, so they could be suspended inside a regular wooden hive. Trying to tie knots in cotton string, while wearing thick gloves sticky with honey and surrounded by thousands of unhappy bees, was an interesting experience; but neither of us got stung, and in Jim's yard the colony survived and performed well. My photographer friend Liz Ball, who luckily had been free that afternoon, recorded the whole adventure on film, and later published an account of it in *The Green Scene*, the magazine of the Pennsylvania Horticultural Society.

I was equipped to act as assistant on that occasion because by then I had my own coverall, veil, and gloves—and my own beehive, parent of the colony I have now. It came about like this: Jim's beekeeping club maintains a "swarm call list" with the local police, a list of members willing to deal with swarms of bees that turn up each spring in people's yards, public buildings, etc., and it so happened that one afternoon, when I was hard at work writing, Jim banged on the door. "I just got a swarm call. Place about five miles from here. Want to come?" His timing couldn't have been more perfect. The teenage characters in my novel-in-progress were themselves about to hive a swarm of bees— something I'd read about and seen done on TV, but never witnessed in person. I turned off the computer and jumped joyfully into Jim's car.

When we arrived at the site, he passed me a hat and veil and a pair of canvas gauntlets, and told me to tuck my jeans into my socks. The swarm, a nice big one about the size of a basketball, hung five feet up in a pine tree. Jim grasped the branch, one hand on either side of the swarm; I, in my borrowed veil and gloves, deep within a cloud of jazzed-up stinging insects, cut it free with a pair of long-handled loppers, then stood back and watched while the colony marched over the sheet and into the hive body Jim had provided.

It was one of the most exciting things I'd ever seen or done, but the best was still to come. When Jim pulled into his driveway afterward, he killed the engine, paused, then turned to me and said, "How would you like to keep this hive and work the bees?" And the following morning the hive body containing the swarm was in my own backyard.

I *could* have started from scratch, with packaged bees bought from a breeder and a very large outlay of start-up money; instead, I've been the recipient of benevolence from several quarters, a really happy way of breaking in. What are the odds of having that kind of luck?

March 31. That will teach me to wax so rhapsodical about bees. I'm typing this with a sore cheek and aching jaw. This afternoon I had just finished working in the asparagus bed, cutting the dead stalks off at ground level and raking away the oak-leaf mulch so the soil could start warming up, when a bee returning to the hive flew full tilt into my face and stung me. I swatted her immediately. The sensation of being stung stopped. But, oh, joy, there was a bee—the same bee? a different one?—caught in my collar or tangled in my hair.

I threw down rake and clippers and ran indecorously back to the house—trapped bee buzzing desperately all the way—pulling off my coat and work gloves, trying to find the bee before I got stung again. Not in my coat. The desperate buzzing continued. I dashed through the house to the bathroom and ran a wide-toothed comb through my hair—and a stricken bee dropped to the floor tiles, obviously the same one that had already stung me, and obviously dying.

I could see the stinger in the mirror, with bee intestines attached, sticking out of my face. I flicked the stinger out with my fingernail, like you're supposed to, and slapped an ice pack on the spot while searching in the medicine cabinet with my free hand for antihistamines, which for some reason we don't seem to have in the house. Then the sting started to hurt like fury. I popped an ibuprofen—luckily, we've got plenty of those—before going back out to finish up.

(When I found the carcass on the floor again that evening and brought it in to show Ted, he peered at it and said, "*Brave* little bee!" in tones of deep solicitude.)

Besides cleaning up the asparagus bed, I picked up the prunings still lying about under the apple trees. I also went around the yard with a bag of Vermont 100 organic fertilizer from two years ago and gave several tablespoonfuls to every perennial plant already established in this yard *except for* the trees, putting fertilizer at the base of the three kiwi vines, around each asparagus-stalk cluster, and on each blueberry bush under the pine-needle mulch. I scratched the stuff into the loose soil; on a windy day like this, the powdery blood meal? manure? in the mix would blow right away if left on the surface.

The "female subtropical kiwi" sent by mistake two years ago, along with the two hardy kiwi vines I'd ordered, isn't supposed to be able to grow this far north. The customer-service person I spoke with at Miller said it would die the first winter, I might just as well throw it away. But that seemed like a poor attitude, so I planted that one closest to the flying buttress that attaches the pergola to the house wall, set a circle of chicken wire around the post, filled it with oak leaves as a sort of gesture at mulching to keep the roots from freezing, and waited to see what would happen.

What happened was that the vine grew like a house afire and has come through two winters in great shape. It's a more vigorous grower than the two Issai hardy kiwis and has much bigger leaves—nice on a pergola. All we need is a male plant and we'd be in the tender-kiwi business, too. Maybe I'll buy it a boyfriend this year, now that it seems likely to survive to reproduce.

I got too busy to train any of the kiwis properly last summer; the vines have all twined around their own stems. Not good. I must sort them out and tie them up properly one day soon, before they break dormancy (which shouldn't be long if we get more days like this one, warm, sunny, windy, high in the fifties).

The buds on the newly planted currant bushes, which were showing green already when they went in the ground, have begun putting out tiny delicate frills: leaves!

Last night I heated five pints of water to just short of boiling, turned off the heat, dumped in a five-pound bag of sugar, and stirred till the whole bagful had dissolved and the syrup had clarified. I filled a quart jar and screwed on a lid with lots of pinprick-sized holes in it. Then I got a coat and a flashlight and took the syrup out to the bees. I didn't want to wait till morning; what if I went out and found a starved-out hive? So I fed them in the dark, by taking off the outer cover and inverting the jar over the oval hole in the inner cover. After the first drops, the syrup won't run out; the bees poke their tongues in the holes to drink. To keep bees from other hives (Jim's, for instance) from sensing the syrup and getting into the frenzy called "robbing," you put an empty hive body over the jar and set the outer cover on top of that. So, for the moment, my hive is three stories tall.

This morning I raised the lid and peeked in. The jar was nearly empty; overnight the hive had consumed almost a quart of 1:1 sugar syrup. This afternoon it was completely empty, with six or eight bees

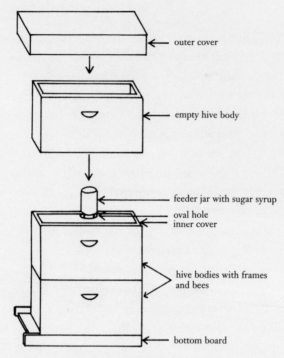

outer cover

empty hive body

feeder jar with sugar syrup
oval hole
inner cover

hive bodies with frames
and bees

bottom board

still sucking at the pinholes. I lifted out the jar and carried it to the back door, where I set it down until the bees should decide to leave. (While waiting I raked the asparagus bed and got stung—a case of biting the hand [cheek] that feeds you.) Later I rinsed out the jar and refilled it from the kettle in the fridge; and when the syrup had warmed up a bit, I took it back to the hive—and though I'd worked thus far without protection, this time I wore my veil.

Just as well, too: when I lifted off the lid a bee flew up and landed on my veil, next to my left eye. Moving the hive into full sun has *not* made this colony noticeably gentler. The bees take little notice when the outer cover is lifted off, letting light in; but the minute the first drops of syrup land on the inner cover—a couple always spill out when the jar is inverted—there's an excited humming and they start coming up through the hole, so thick that it's hard to set the jar down without squashing any (and I usually do).

On the seedling front there are minor difficulties. In the two years since I last started seedlings, I'd forgotten the reasons why bottom watering is better than top watering and am paying for that mistake. Two eggplant seedlings are dying from damping-off, a constriction of the cells of the stem at the soil line. And I think I know why the rest of the Gurney Girl tomatoes aren't sprouting. The particles of peat and vermiculite in Jiffy Mix don't stick to each other; the mix shifts like sand when watered from above. For all my cavalier comments re: not fussing too much about covering the seeds twice as deep as their thickness, my too-casual top watering has made the mix drift in the pots. The seeds may now be too deep to germinate, even with bottom heat. If they are, I'm going to have to start over.

April

April 1. Rain was forecast for this afternoon, so this morning I went out and pruned and tied up the kiwi vines. (I'm beginning to think the best way to *ensure* something gets done is to mention in this record that it needs doing.)

There was a good article on growing kiwis in *National Gardening* a couple of years ago. I saved that issue and studied it during the first two seasons of my vines' growth (though today, wouldn't you know, I can't find it anywhere), and I think I understand the structure of a kiwi vine—main stem, cordons, fruiting laterals—and where each of them should go on the trellis (or, in our case, pergola). I even understand how I *ought* to have trained my kiwis last summer to produce three pergola-shaped vines. That classic shape may now be beyond my powers to impose on them without sacrificing the potential that they might bear this year.

Be that as it may, I pruned the tangled tentacles back to green wood and a big bud, sorted out all the strands, and tied them neatly to the pergola rafters with stretchy green acrylic tape, one vine to a rafter. The various ties from other years, strips of ragbag scroungings and rusted-out twistems, have all been dumped in the trash. I cut off one cordon at the main stem of one of the hardy kiwis, not realizing—it was hard to see which tentacles were which, they were so intertwined—how much of the top growth originated there. A mistake, but not a fatal one.

The tender kiwi had some kind of damage to the bark on one of its main tendrils. Frost damage? Disease? Whatever, I cut that one off at the base. The vine will make more; it's incredibly vigorous.

So that's one job well done. And it *was* done well—in a good, unhurried, calm spirit, the spirit I want to cultivate for the whole Homestead Year. The stepladder is a rickety old thing, and the pergola made an unsteady base on which to lay my tools—clippers, scissors, roll of tape, roll of plastic-covered wire. I sometimes had to stand on the very top of the ladder, my arms and torso sticking up between pergola rafters, in order to work. All my movements had to be calm and careful, and they were; I didn't drop anything and I didn't slip or wear myself out with the physical tension the task might easily have generated.

Later, safe in gloves and veil, I fetched the now-empty, bee-encrusted jar back from the hive and refilled and replaced it. The first five-pound bag's worth of syrup is nearly gone. The books say: once you've started feeding, DON'T STOP—the bees are raising brood and are relying on the supply they've grown accustomed to, to feed a lot of hungry larval mouths.

Oh me. Even though I got so much essential equipment as hand-me-downs, I have to admit that this hive has cost an impressive amount of money. On the other hand, how do you put a price on the experience of observing the bees through an entire year? Anyway, once we take some honey off, the expense will seem unimportant. I only fret because so far we haven't reaped a drop.

At the first opportunity, i.e. a calm, sunny, 70° day when I'm home, I need to take the hive completely apart and see what's what in there.

My cheek was more swollen when I got up this morning than it was last night when I went to bed. It hurts less, but has started to itch. But I'm not sorry I got stung; now I know I'm not violently allergic to bee stings. If I were, this sting on top of the one last August would have had a much more serious effect.

I restarted the six Gurney Girl tomatoes that never came up and put them back on the hot-water heater. Some of the lettuces and cabbages have true leaves now.

Miserable slow spring! Cold and cloudy and windy, day after day after day.

April 4. Today I opened the new twenty-pot tray and started more seeds: Red Sails lettuce X 4, Black-Seeded Simpson lettuce X 4, Tam mild jalapeño peppers X 4, and 8 pots of pennyroyal seed, tinier than grains of salt, which I sprinkled over the surface of the moistened potting soil and pressed in—no way of measuring twice the thickness of *those* seeds! The packet says they need light and constant moisture to germinate, so straight under lights they went. The others, on plastic plates in plastic bags, have joined the restarted Gurney Girl tomatoes on the hot-water heater.

As for the Gurney Girls, still nothing doing there; but I feel better about the possibility that I may have caused the problem by top watering. I looked up the records on Gurney Girls. In 1989 I sowed the seed on March 17, and they were all up after twelve days; but in 1990 all three tomato varieties I was growing were slow to germinate. I'd started ten Gurney Girls on March 20; on April 9 I wrote: "A new sprout up today, but these germinated poorly—presprouting some seeds." On the 17th: "presprouted seeds *with seed leaves* survived if root planted in soil and seed leaves left exposed—if seed leaves buried, they never emerged; ditto seeds w/out seed leaves * important discovery." Finally, on April 19: "several more sprouted, a full month after planting!" The record later states that even the youngest sprouts were big enough to plant out by May 18, and that all grew to be immense and were very productive.

Two lessons here: (1) a late start is a problem only if you're trying to have the first ripe tomato in town; (2) it pays to keep records. I'm not going to fret about it anymore.

Pennyroyal, a member of the mint family, is said to be a good natural insecticide against the deer tick. Since we've got Lyme disease all over the place hereabouts, and since I don't want to spend my summers smeared with DEET or sweltering in long pants and long sleeves, it seemed a sensible thing to grow—though you start the plants indoors in early spring and don't plant them outside till *late* summer, so obviously they take a while to get going. But, once established, they're perennial; I won't need to do this again.

The last jar of syrup has been fed to the bees. Unless the weather warms up right away—enough for me to make the thorough inspection I need to make—I'll have to melt another bag of sugar and keep doing this.

April 5. A sunny, fiftyish, very windy day. While sorting and tidying in my study this morning, I found the issue of *Organic Gardening* from the summer of 1986 with that article I remember reading, about gooseberries and the bad press they, like all the *Ribes*, have had on account of white-pine blister rust. I skimmed the article in hopes of finding that it was all a mistake, that black currants at least were ever innocent of all wrongdoing. But the author states that the alternate host of the disease was, not the maligned gooseberry, but certain native *Ribes* species *and European black currants.*

So—is Prince Consort a *European* black currant? Or a European or native black currant with resistance to being an alternate host bred into it? Or what? I'm beginning to get the feeling that nobody really *knows* for sure which *Ribes* species pose a hazard to white pines. I should keep an eye out for anything published on this subject.

For now, though, I'm sticking by my three little bushes, with all Mr. Miller's authority behind me.

Three chicken-wire tomato cages, held up by stakes threaded through the wire, have been standing in the garden all winter. Inside were the dried remains of three volunteer tomato plants that sprang up and flourished in three different beds, in the gardenless summer of 1991. Today I wiggled the stakes loose and pulled them out, and broke the dead vines off at ground level—a little piece of housekeeping that made the garden look much tidier.

We haven't yet woodchuck-proofed the garden fence, but there are two crops that need to be planted pretty soon: onions and potatoes. Woodchucks don't bother onions (famous last words), and the potatoes will be under mulch for a while, so it should be okay to go ahead and plant. This afternoon I went out with ground cloth, rakes, and hoe, and tried to move the year's accumulation of dead leaves from the end of the garden where they've spent the winter—which is also the site of the future potato patch—to another holding area, till such time as they're needed to mulch paths and be mixed with grass clippings in the compost. The leaves are all matted down and wet just under the surface; a hoe is the best tool for chopping chunks loose from the mass, and a garden rake the best tool for moving the chunks onto the dropcloth. But once they've been broken loose, the dry leaves at the surface can blow away—and this was a *very* windy day, with gusts of twenty-five mph. The wind pushed my dropcloth around and blew leaves all over the place. After a while I realized it wasn't a good day for the project planned and abandoned it for something less frustrating: digging one of the growing beds, the first of the season.

I was pleased at the way I handled the shift from leaf mining to digging. It would have been easy, and quite in character for me, to have gotten furious at the wind for disrupting my plans, and/or to have persisted in trying to move leaves in spite of the sort of day it was.

The bed I dug is to contain onions. I racked my brain, but could not remember where onions had grown the last two times, and in the end had to go inside, walking on the edges of my muddy shoes, and consult my garden records in order to pick a bed where no onions (not counting onion grass) had grown for at least three years. One of the organic gardener's main lines of defense against pests and diseases is to rotate crops, so that bugs and microorganisms which overwinter in the soil—downy mildew and smut fungi in this case—wake up next to something other than their favorite foods.

These raised beds have never been walked on. The narrow permanent paths between them run north and south inside the fence. Some beds are bigger than others; the one I spaded up today is about three shovels wide and twenty feet long. Our heavy, yellow clay soil has absorbed tons and tons of organic matter in the years we've lived here; and that, plus not being walked on and compacted, has made the soil in the beds far looser and fluffier—and browner!—than what comes out of the path if the spade cuts wide of the bed's edge by an inch.

But none of the beds has ever been double-dug, either. If I can get to it, I'd like to double-dig *one* bed at least this summer, or maybe half of one, as an experiment to see how much difference it really makes. Double-digging is a lot of work. You dig down a foot, removing the soil, then loosen the subsoil below with a pitchfork to the depth of another foot, theoretically making it much easier for plant roots to grow deep and spread far. This, in turn, produces lush top growth and great yields. But our yields are pretty good anyway, so it's one of the projects I keep not following up on.

This coming week, I must try to collect soil samples and prepare the soil test to send to Penn State.

The bees have polished off the third batch of syrup. When I went to take the jar out of the hive, I watched to see if they were bringing any pollen in, and was surprised to see that a lot of them were—a whitish pollen, from who-knows-what in bloom. With the temperature right at fifty and twenty-five-mile-an-hour winds, I wouldn't have expected them to be able to do any foraging to speak of, but evidently they can.

Every time I remove the empty syrup jar, there are a dozen bees, more or less, still clinging to the lid, trying to lick up the last vestiges of sweetness. They won't let go. I turn the jar upright and set it down on the ground, close to the entrance so they can make it back inside despite the cold. But a couple of nights ago when I went out after dark to get the jar, two or three bees were still on the lid. When I pushed them off with one gloved finger, they fell to the ground, dead; they had died suckling at the pinholes, refusing to give up.

On my walk around the neighborhood this afternoon I saw a big flock of robins poking about in the dead leaves of a wooded lot. Robins have been overwintering here in recent years; but, come to think of it, I don't remember seeing any this winter at all. In any event they're back—not pairing up as yet, but getting ready.

One of the perks of living in the Borough of Rose Valley, which I've been told over and over has the highest per-capita income in the Greater Philadelphia area, is the three-mile loop we've discovered/invented for our walks. The loop runs through back streets and roads and

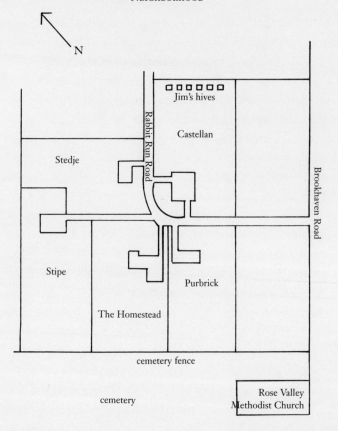

one private yard, crosses two creeks by little bridges, skirts the tennis courts and local swim club, and for half a mile follows a densely wooded path—formerly a trolley right-of-way—where tracks once connected Swarthmore with Media (our local bank-and-shopping mecca), a distance of some five miles. From there it crosses the tracks at the commuter rail station and swings past a post office and a convenience store before turning back toward Rabbit Run.

For Ted and me, a typical teaching day begins with the walk to the Moylan–Rose Valley commuter rail station—a walk which happens to coincide with the first mile of the three-mile loop, and which is un-failingly delightful in all but the worst weather. At the station we hop a train, the Media Local. In a little over half an hour it will deliver us to 30th Street Station in Philadelphia—one of the many commuter trains flowing into the station's beautifully renovated bowels out of the northern and western suburbs and New Jersey, bringing people to work in the city.

Rose Valley, in Delaware County, is fifteen miles southwest of 30th Street. For a few of those miles the view from the train window is lovely: woods, the Swarthmore College campus, woods again. But as we approach Philadelphia this is less and less true; and before we reach the station, the tracks will run through scenes of poverty and squalor as dis-mal as any to be seen on the eastern seaboard. Our commute ends with a fifteen-minute walk to the university through a concrete landscape unrelieved by so much as a blade of green.

After the long day in the Halls of Ivy, the journey is repeated in reverse. Somewhere on the trolley path we manage to shuck off the city. But never the awareness of our extraordinary privilege.

April 7. Another sting! I was moving more leaves, working with rake and hoe, aware that bees were zipping by, when bam! another one hit my head, got entangled in my hair; I tried to bash it before it could sting me, but no luck. When I went inside and looked, there was the little back stinger embedded in my scalp.

The hive is in the wrong place; that's pretty clear now. In its former position, under the trees back by the back fence, it never gave this sort of trouble; but I'd read that bees kept in the shade seem to be more aggressive, so during the cold weather, while they weren't flying, Ted and Jim and I moved them into the sun. Now what? They can't be moved now—not more than a foot or so per day—or they won't be able to find their way home.

I'd planned to do the serious spring inspection this afternoon if it warmed up enough; but by then I'd been stung, and didn't want to open the hive smelling of sting pheromones, which would rile up the bees. Tomorrow without fail. An article in this month's *Gleanings in Bee Culture* warns that steps should be taken the first week of April, on average, to prevent swarming by April 20. The author tells what to look for. I'll spend tomorrow morning planning my strategy and open the hive right after lunch. Then we'll see.

They're working hard, the few days they're able to get out. Today's sixtyish sunny-cloudy weather had them zipping around like bullets. I went and sat in the bushes and watched to see if they were bringing in pollen. The pollen baskets of some were so packed that you could see them even in flight, like bright yellow hubcaps.

Other developments: last night when I got home from town I went down to see what was what with the seeds. All the lettuces I started on Saturday the 5th were up, but not the peppers, and not—still not—the Gurney Girls. No change this morning, so I took the last four seeds out of the packet, wrapped them in a wet paper towel, and stuck the towel in a plastic bread bag; I'll keep them warm and watch them for a week. If they don't germinate, it'll be time to throw in the towel. Whatever happens, I'm afraid the main tomato crop is going to come in late. We should have plenty of tomatoes for salads from the other plants, but spaghetti-sauce production will be slowed down, with the risk that it'll run right up against Labor Day and the hectic first weeks of the fall semester, right when the applesauce apples will also be rumbling in. Ted usually does most of the sauce processing; but this year I expect to have to pull more weight than usual myself.

Speaking of applesauce, though: the Winesap boughs in the vase on the hearth are well leafed out now, with not a sign of a single flower. I conclude that once again we will get no Winesaps this season.

After getting stung I gave up leaf mining, pending a solution to the problem. Instead I went and opened a furrow in the dug bed, which is safely out of the flyways, dumped in bucket after bucket of overwintered compost from the big can, then closed the furrow and worked quickly up one side of the bed and down the other, spading the compost into the soil. That done, I raked the whole bed into the shape of a very long meat loaf, smooth and round—flattened the top—and planted the onion sets four inches apart each way, pushing the little bulblets down into the soil. Finished, the whole planted bed is dimpled with a pattern of little push-marks. All the while my head was burning in that cold bee-venom way, but again, it wasn't a bad sting. But worrisome: I will have to deal with this, solve the problem. I don't think it can be worked around. And a solution needs to be found pretty soon because it's getting to be time to plant potatoes.

I found a couple of little round undug potatoes from last year in the onion bed and saved them. You're warned against letting these grow, since they might harbor disease, but I haven't had any trouble of that kind. We get volunteer potatoes every year; soon they will have been rotated through so many beds that every single bed will grow potatoes, even if I don't plant a one. It seems impossible to find them all at harvest time; but this year may be an exception since they'll be grown under mulch, hence unable to hide in the soil.

When I'd finished planting onions it was after four, the hive was in shadow, and the bees had calmed down. I moved the rest of the compacted leaves off one of the four-by-eight foot boxed beds in the site of the future potato patch, so I could dig around the inside of the box, which is going into its next incarnation elsewhere as a worm farm. This box was made several years ago of untreated two-by-eights and has rotted somewhat from many months of contact with soil moisture, but it'll hold together a while longer—long enough to raise a lot of red wigglers to feed the catfish scheduled to be fattened for the table, in the pond we mean to dig this summer.

One of the great mysteries about this yard that grows so many vegetables so well, and produces such large numbers of white grubs, is: why is there never a single earthworm? The soil is loaded with organic material. Where are the worms? One of the boxed-bed frames will grow red worms for the fish—red worms reproduce faster, so they say—but the other is to be an experimental growing tank for ordinary garden earthworms. I plan to fill it with soil from the fishpond excavation, mixed with leaves etc., and import worms from neighbors who have lots of them. If they thrive, the problem's not our soil, and I'll start introducing worms into the garden. If they die . . . well, then I guess the problem probably *is* our soil, and then we get to decide whether to have it expensively tested for contaminants.

Sunday and today, I just realized, I not only saw no worms when I was digging the onion bed, I also saw no white grubs. Usually there are *tons* of those, gross things, curled upon themselves, waiting to eat the root hairs off my seedlings. This time, not one. The county extension agent told me the earthworm problem couldn't be contaminated soil if the soil supported white grubs. So now what?

Before calling it a day, I went around with a shovel and a plastic dishpan and collected some seventeen samples of soil from six to eight inches below the surface, to mix together and send away to be tested; but the test they do at Penn State is for pH and nutrients, not contaminants. The people who owned this house before we did certainly never put chemicals on the soil; the lawn as we've known it has always been a community of dozens of plants, more of them weeds than grasses. Testing for things like lead and pesticides costs a bundle. For the time being, these remain open questions.

April 8. A traumatic day at the hives. I devoted the forenoon to a study of my many bee books, trying to form a plan for the afternoon's spring inspection. Between keeping the smoker going, bees roaring around, being handicapped by the veil and too-large leather gloves, and trying to handle too many tools under these clumsy conditions, it's hard to think clearly while the operation is under way (especially right after having been stung). So I read up on the subject and made a list of things to look for. Then I made a second list—of things to take with me—so I wouldn't forget a crucial piece of equipment.

After lunch I dressed for the encounter with the bees as carefully as a matador before a bullfight, leaving nothing to chance; I did *not* want to get stung again. It was a beautiful bright day, the warmest of the spring, nearly 70; my bee coverall over a sweater was hot, but I wanted lots of layers. I wore my black rubber Wellies and tied my bee veil down tight. And off I stalked, smoker smoking, rendered semiblind by

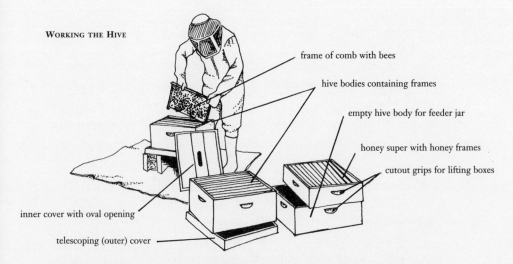

WORKING THE HIVE

frame of comb with bees

hive bodies containing frames

empty hive body for feeder jar

honey super with honey frames

cutout grips for lifting boxes

inner cover with oval opening

telescoping (outer) cover

the combination of my reading glasses (for close work) and the veil, toward the other side of the yard, where things were definitely popping.

A typical modern beehive consists of two wooden boxes without attached tops or bottoms, set on top of one another. Each box, called a *hive body*, or sometimes a *deep* or *full-depth super*, is exactly the right size to hold ten wooden frames of the wax foundation sheets on which the bees *draw* (build) the honeycomb. (Nine frames are sometimes preferable, and spacers are available to keep the nine frames suspended evenly in a box designed to hold ten.) As I've mentioned, these movable frames hang in the hive body like hanging files, suspended from strips at the rims. The stacked hive bodies are set on a *bottom board* providing an entrance at the front only, and closed at the top by a thin *inner cover* with an oval hole in the center and a small *half-moon port* on the rim, and over that a much sturdier, lidlike, metal-clad *outer*, or *telescoping, cover*.

Two hive bodies with nine or ten frames each provide enough space for a colony of bees over the winter; but in spring their numbers increase dramatically, up to forty or fifty thousand, and the two-story hive can begin to feel crowded. The worker bees respond by making another queen, or sometimes several queens. Then, just before the new queens emerge, the old queen leaves with about half the colony to find new quarters and make more space in the old hive. When that happens, the hive is said to be swarming.

The absconding bees alight someplace near their old home and form a ball, or cluster, around the queen. They may stay in the cluster for half an hour or a day, depending on how long it takes their scouts to locate a suitable second home. It's during this period that swarm calls come in to the police from agitated homeowners and are relayed to people like Jim, who are happy for the chance to acquire a free hive for the cost and bother of removing it from where it isn't wanted. All they have to do is persuade the swarm that the hive body being offered to them is the "suitable place" they've been looking for.

In nature, swarming is a fine strategy, serving the needs of the bees very well. But if a beekeeper like me wants surplus honey, to serve her own needs, she must try to prevent swarming; fewer bees mean a weaker hive and less honey at the end of the season. So the main purpose of the spring inspection is to find out whether it looks like the hive is making preparations to swarm and, if so, to take preventive measures. The other purposes are more prosaic: to see if the hive has enough stored food to last until the main honeyflow (season of nectar-producing flowers) begins—I already knew mine didn't—and to find out, from the amount of eggs and "capped" and "uncapped" brood (larvae) present, whether the queen is still laying well. An egg, deposited in a cell of the honeycomb, hatches in three days; the larva is fed for six days; then workers "cap" the cell. After pupating, the larva emerges twelve days later as a new worker bee. One of the things I hoped to find today was a lot of frames containing solid expanses of capped brood, cells covered with a brownish wax under which larvae are turning into adult bees—a sign of a fertile and productive queen.

I can recognize capped brood, all right, but for important inspections like this, my lack of experience is a major handicap. So much else is going on in the hive, most of it on such a small scale, and it's so hard at first to understand what you're seeing. It's also the case that what I actually do see, more often than not, isn't what I expected to see (and this time was no exception).

Thinking out my strategy in advance was helpful. I smoked with restraint. I remembered to move slowly and to check for the queen before putting the telescoping cover down in one place and the inner cover in another, so as to have two surfaces on which to place the two hive bodies—I was going to tear the entire hive apart, clear down to the bottom board—"down to its toenails" as the veteran beekeepers say.

First to come off was the empty hive body protecting the sugar syrup from vagrant alien bees. Then the inner cover. Then I puffed in more smoke. The top box was crammed with bees, all upset. Still moving slowly, I began unsticking the frames one by one with my hive tool and pulling them loose with my frame gripper. The frames were so thick in agitated bees it was hard to see anything, but I located some honey and a little capped brood. What freaked me out, though, was the presence of some lumpy objects on the bottom bars of the frames.

I saw those on frame after frame and panicked. When a hive is preparing to swarm, inch-long pendulous "queen cells" are built of wax on the bottom bars by worker bees, and the larvae inside are fed on royal jelly until time to cap the cells. As long as the queen cells are open, you've still got time to act; but book after book has assured me that once they're sealed over the hive is committed to swarming and the swarm is unstoppable. I've never seen a ripe queen cell—only drawings—but these lumpy things looked more like queen cells than anything else I could think of, and they were definitely capped. I was positive the little ingrates had made up their minds to swarm.

To try to stop the process against the odds, some experts advise cutting all the queen cells you see; but they also warn that this may leave the hive queenless, since the swarm will probably leave anyway, taking the old queen with it. I compromised: I cut one cell and left the rest—achieving precisely nothing.

Alarmed—to put it mildly—by the prospect of having to try to prevent a swarm (by making the hive think it *has* swarmed—there are several ways of doing this), I continued to inspect both sides of every frame in the top box as best I could. There was a tremendous lot of capped brood. Some of it, not enough to be concerned about, was drone comb, which looks like the protruding noses of bullets in a box (capped worker brood is flattish). Workers are infertile females; drones are fertile males whose job is to inseminate queen bees of their own and other hives. Since they do no other work and have big appetites, you don't want too many drones in a hive.

But I was trying to identify a balance between eggs, larvae, and capped brood, which would show that the queen had been laying right along in a consistent way. And look how I might, I could see no larvae and no eggs, and was swept by a fresh panic: the queen must be dead! No queen, and a swarm imminent! Bad news—the worst!

A full frame of brood is both heavy and delicate; it's practically impossible to shift one out of the frame grip, so that both hands can be used to hold and turn it, without damaging the thing or dropping it, or squashing lots of bees. Sweating with the strain of concentration, I heaved up the top box when I'd finally finished sorting through it and set it down, with all its bees, on the outer cover. They were such unhappy bees that I decided to smoke them again—and that's when I discovered that my smoker had gone out.

There was some capped brood in the bottom box, but no eggs and no little white comma-shaped larvae curled at the bottom of the cells of comb. Or none that I could see. Having my smoker poop out on me made me hurry a bit more than I'd intended to.

I set this bottom hive body on the inner cover, exposing the bottom board to the light of day for the first time in a year. I'd expected to find dead bees piled up behind where the entrance reducer had been, and had brought my bee brush to sweep them out, but there were none at all. The bottom board was clean, except for some little drifted pills of wax or propolis crumbs in the corners. Sweeping *them* out seemed not worth the bother of disturbing all the bees on the board, so at that point I just started putting the hive back together.

The conditions I'd found weren't the ones my reading had led me to expect; but since you do the spring inspection to find out what conditions in the hive are actually like, I couldn't complain. But also I wasn't at all sure what action to take and was too exhausted and worried to think clearly. Obviously, I should do something to stop the hive from swarming, but what? This morning I'd read—and even understood (I think)—two articles, each describing a different method of swarm prevention. But both require several frames of *uncapped* larvae, and I didn't seem to have any at all. Then I thought of something. As part of their regular spring management program, a lot of beekeepers reverse the top and bottom hive bodies so the mostly empty (by then) bottom box gets placed on top of the mostly full (by then) top one. They do this because the queen, and therefore the brood nest—the area of comb where she's currently laying—tend to move upward in the hive. The top box is mostly full in spring because she's been working her way up all winter. Reverse the boxes and she's back in the bottom with space above her, feeling less crowded and therefore less motivated to swarm.

I'd read two directly contradictory articles about reversing when there's a brood in both boxes, one saying it's a great idea and one warning against it. Actually, at that moment I forgot the arguments against; I was driven by a sense of desperate urgency, as if the swarm might take off in minutes if I didn't do *something*. So I heaved up the top box, fifty pounds if it was an ounce, and set it down as gently as I could on the bottom board; it was now to be the bottom hive body. The *former* bottom box, which was relatively light, I set atop the other one. Then the inner cover, and then the outer.

I dragged all my equipment back to the house and climbed out of my layers of protective clothing. The next several hours were spent deep in bee books, hoping (in vain) that somebody would have described my situation. Finally I had the sense to go find the "queen cell" I'd cut off and pitched into the undergrowth. Less manic dissection revealed large larvae and milk-white bee shapes larger than workers, but the undamaged caps were bullet-nosed. This was no queen cell, it was a wad of burr comb—lumpy comb built anyplace other than on the wax foundation—containing a number of drones at various stages of development. Drones, not queens.

Then another thought struck me. Workers in a hive about to swarm stop foraging and hang about the entrance; my worker bees, by all-too-obvious contrast, were flying around like gangbusters, stinging people in the way of their purpose. At this point I began to calm down and feel embarrassed. There was, however, still the question of the missing eggs and larvae, so this evening I made a hasty phone call to Jim Castellan and described what I'd seen.

Jim relieved my mind immensely by saying he'd encountered the problem once himself, had had the same wild reaction—The queen's dead!—but had been told by an old hand not to panic, that the queen sometimes takes a break between cycles of laying. And that seemed to be the case, he said, because everything had come out fine. Give it a week or so and check again.

All this fussing took up most of the day and all my strength. I did find a few minutes to thin the lettuce and cabbage seedlings to one seedling per pot. On the verge of doing the same to the tomatoes, I suddenly decided to throw in the towel about the Gurney Girls and transplant some of the redundant Beefmaster and Viva Italia seedlings into the still-empty Gurney Girl pots.

I also opened the bag of seed potatoes, thinking to let them start presprouting. Too late: every potato had grown at least one long fleshy tentacle, and most had a lot of them; it was hard getting them out of the bag without breaking the sprouts off. I cut the spuds very carefully into pieces and set the pieces in cardboard boxes to form calluses and keep sprouting. Friday Ted thinks he'll have time to help me with a potentially ingenious idea I have for bee-proofing the garden, so I can get the potato patch ready to plant.

April 9. The morning was mild and sunny, springlike. I cooked up a new batch of syrup, using the second five-pound bag of sugar, and medicated it with Fumadil-B according to the instructions. Fumadil is a nosema (dysentery) drug. I also mixed Terramycin into two ounces of powdered sugar; that's for American foulbrood, the most serious bee disease I have to worry about.

When these medicaments had been concocted, I carried my

basket of beekeeping tools out to the patio and spent an hour or so cleaning up one of the honey supers bequeathed to me by Jeff and Liz Ball, using my hive tool to scrape the hardened propolis from the wooden frames. When I'd cleaned all ten frames and the super itself, I got two nine-frame spacers out of the box of new equipment from Kelley, the bee-supply company, figured out which nails were the right ones and how the spacers had to fit in the super, and briskly nailed them in place. Nine nice clean frames are now hanging in the nice clean super. Most beekeepers have shifted over to nine. A hive body containing ten frames is hard to work in oversized gloves, what with propolis gucking everything up. And you get just as much honey, because the bees fill the extra space between frames by drawing out the comb a bit farther—which makes uncapping at extracting time easier, too.

The honey super is a box of frames identical in every way to a hive body except for being shallower—$5^{11}/_{16}$ inches versus $9^{1}/_{2}$ inches. This one is for giving the bees more headroom now—a swarm-preventing maneuver (I hope)—and then eventually for surplus honey storage. *Super* is Latin for *above*; I've always assumed shallow supers were called supers because they're put on top of hives—except that an ordinary hive body is also sometimes referred to as a full-depth super. Go figure.

Anyway, I was able to proceed in a steady, methodical way, and the work consumed most of the morning and was very enjoyable.

Then I suited up and fired up the smoker. The day had clouded over and become windy, and the bees were flying less zestfully; I felt bad about disrupting their lives again so soon and exposing them to the chilly wind, but I worked fast and everything went smoothly. After plunking the super on top of the hive, I sprinkled the antibiotic-laced powdered sugar on the ends of the frames' top bars, which is where you're supposed to put it, God knows why. The inner cover went over that, then the jar of medicated syrup inside its empty hive body, then the outer cover. Voilà.

I'll leave them alone now for a week or so, except for feeding and medicating. With luck, when I check again there'll be eggs and uncapped brood in there, and that will give me the means to try to thwart a swarm.

April 10. Ted and I spent the morning devising—and the early afternoon putting up—an ingenious device for making the bees fly higher over the garden; and, except for a few nuts and bolts, and twenty feet of nylon cord, we did it all with stuff we already had around.

The materials we used were two lengths of shade cloth five feet wide, used for the past two years to cover the pergola (which gets ferocious afternoon sun in summer), three eight-foot poles cut from the jungle along the cemetery fence and trimmed of branches, some wire, and the aforementioned nuts, bolts, and cord.

We needed a way to force the bees to change their flight path. They'd been zipping back and forth over the garden fence like shrapnel, right at neck height. I wanted to be able to stand in the potato patch and have them clear my head by a good two feet. My idea: to create a barrier by attaching the shade cloth to the poles and raising it above the three-foot garden fence, securing the poles to the three equidistant metal fence stakes with wire. Neither piece of shade cloth was long enough by itself; together they made a strip much *too* long for the purpose, but I didn't want to cut them. We solved that problem by fastening the two pieces together and rolling the outer ends around the end poles.

The result resembles a long scroll of green netting with a seam up the middle. The two pieces were joined by stapling each cut end tight to a lath and nailing the two laths together. We then bored holes through both laths and the center pole to receive the bolts.

We've ended up with something that looks like a very tall green volleyball net set up on three posts instead of two. Ted—ducking bees—held the center pole tight against the center fence stake while I, wearing my bee helmet and veil, twisted wire around both pole and stake, and tightened the wire with pliers. Then we raised and wired the end poles, first one, then the other. The netting drooped and flopped around in the wind, but for the moment it was up. Later I got the stepladder and bunched and tied the scroll ends at the top with the nylon cord.

It was a brilliant, sunny day, the warmest yet this spring. We stood admiring our handiwork and watched the bees zip straight into the netting, bopping into it both coming and going from the hive. With the sun glinting on their wings we could see them very well. I'd hoped the dark green shade cloth would be visible to them, but they couldn't seem to see it. I suppose these are field bees that have been going and coming on this flight path for a while now, and simply couldn't believe that this large obstacle had sprung up suddenly to block the way. Now I'm thinking, if I can just tough it out for a couple of weeks, this whole generation of field bees will work itself to death, and the next generation will encounter the barrier on its first flight, and learn to avoid it.

Thunderstorms predicted for tonight. Will the netting, which whips around in the breeze pretty good, allow enough wind through so that it doesn't rip loose? Let's hope so.

When I first tried the idea of the Homestead Year out on Ted, he was cautiously supportive but made it perfectly clear that this was my project, not his, and he was not to be viewed as being an equal partner in the undertaking, or automatically on call when I got into difficulties. Fair enough; but I know Ted, and that's the stance he takes toward any new undertaking if it strikes him as reckless or extreme. Or takes initially. Once he sees that I've accepted his reservations and really don't intend to pressure him to do anything he doesn't want to do, he starts sniffing around. The next thing I know, he's pitching in with increasing enthusiasm; and pretty soon, if I don't watch out, he's trying to take over.

Where the homestead is concerned this is a very good thing, because a project like this bee barrier would be harder by a factor of four or five to bring off without a working partner. Not impossible, just harder and much more frustrating.

In a continuing spirit of helpfulness Ted helped me move the bed-boxing boards (which came apart when we tried to lift the box— rotted out; they'll have to be refastened some way when the worm farms are established).

By late afternoon the day had clouded up, so the bees calmed down and I went out and spaded up Boxed Bed #1 (sans box), pulling out thistle runners and digging in four or five pailfuls of compost, which isn't nearly enough for optimal results, but is all I thought I'd better put in—there's lots more garden for the four remaining canfuls of compost to be added to. A bee or two wandered over and poked around in the moist soil, but none crashed into me. The bed looked very luscious when I'd raked it smooth. Tomorrow I'll do a soil test on the sample for pH and fertility.

Funny how the bees don't seem at all interested in forsythia. This is a great forsythia year—boughs covered with yellow stars all over Rose Valley, no leaves at all as yet. Forsythia flowers must have no usable nectar, but it seems a terrible waste.

April 11. To my amazement, the bees appear to have learned *already* to fly up and over the barrier! They've stopped bumping into it, at least the ones we saw this afternoon when we went out to watch them. The netting's still standing, but the forecast storm never developed, so there's yet to be an acid test. I wonder what the neighbors are saying. No, I know: *"What are they doing now? What is that thing?"*

Let it be said, though, that Rose Valley, despite (because of?) its poshness, is actually the most noninterfering community I've ever lived in. Since households here have an acre apiece to conjure with, laissez-faire may be the simplest way to run the place. Ted, who lived for many years in Swarthmore, says it's the same there. Take lawns, for instance. Ted still remembers his relief at coming to Pennsylvania after years in New Haven, where his neighbors fussed and fretted continually over their lawns and expected him to do the same. With a couple of exceptions, scruffy lawns full of weeds are the norm at our end of Rabbit Run. Grass does get mowed, but not fertilized or watered. Ted throws compost around in summer and woodstove ashes in winter; others may do less.

There *are* exceptions: people who devote tremendous amounts of time, money, and work to maintaining their large yards in a beautifully groomed state. I know it gives them a lot of satisfaction. My personal sympathies, however, have always gone to the likes of the Castellans, who used to hang the washing in the *front* yard—perfectly sensible, the rest of their property is too shady for anything to dry—and whose place always looks like a lumberyard, for the very good reason that a glowing woodstove in the family room is the sole source of heat for their extremely attractive house.

I'm sure there are neighbors who wish that it were otherwise, but wishing is as far as they will go. And it'll be the same with the homestead. Where else in typical suburbia could I get away with raising ducks? (Swarthmore, says Ted.) Though I deliberately haven't checked, I *know* our zoning laws proscribe poultry; I also know, however, that if nobody complains we'll be left alone. It's the Rose Valley way to live and let live, so long as nobody organizes a drug ring or a motorcycle gang or something *truly* antisocial in the borough.

I ran the soil test, using a Sudbury kit from a local nursery. The pH is very low, around 5, but that's okay for potatoes. The nitrogen reading is quite high, probably thanks to the compost. Phosphorus is low, potassium is so-so. No problem, I can adjust the fertilizer mix to correct the balance.

The magnolias have split their britches, the pachysandra is flowering, and so, finally, are the daffodils, everywhere. But I still don't know what that yellow pollen is that the bees are bringing in.

Down cellar, the biggest cabbage seedlings look terrific, and the biggest tomatoes are perking along. Lettuce in the seedling stage is so fragile you'd swear it'll never survive, but it always does. Eggplants are finally each growing their first true leaf. Peppers still not up—it's probably just too cold as yet down there for peppers to germinate. As for the Gurney Girls, they're clearly a total bust.

April 12. This afternoon, after a long, dreary morning of taxes, I rushed gratefully outside to dig the second boxed bed in preparation for planting potatoes. The boards were just as rotted as those surrounding the other bed. I schlepped them one by one back to the jungle before setting to work.

It was obvious at once that I was up against a very different proposition in this second bed, which has had everbearing strawberries in it for the past two years. Even without the problem of thistles invading the bed, the strawberries were not a success—pretty but flavorless. The Canada thistle made it easier to decide to dig the whole mess under, but the thickly tangled strawberry roots and two years of not being turned made the soil dense and compacted, and I had to work hard to spade it up. But the job got done, finally. I spread another four or five buckets of compost around and raked it into the surface soil enough to keep it from blowing away if things get windy. In a couple of days I'll turn the whole thing again.

The soil thermometer in the other bed reads 50° at about seven inches, 54° at two inches, 55° at one inch. Getting there. I know from experience that unless the soil is warm enough—65° or thereabouts—potatoes won't root, they just sit there and sulk.

Ted saved me a couple of buckets of sawdust that he'd produced while cutting stovewood, and I used them to mulch the currant bushes, which I also watered. We haven't had more than a trace of rain this whole month; the clay was deeply cracked right across the planting holes.

The apple trees are breaking dormancy now, with little rosettes of leaves where the blossoms will be (not on the Winesaps, though). The Winesap wands have today been replaced by forsythia branches in the tall vase on the hearth.

April 14. Spring's finally here. This has been a great day, cool and sunny, wonderful for working outside. I mixed up a big batch of fertilizer, using the formula from Mel Bartholomew's book *Square-Foot Gardening*, that I've been fertilizing plants with since I first started. This batch has been amended to double the potassium content: four parts compost, *six* parts wood ashes, two parts bonemeal, and one part blood meal, to give NPK values—nitrogen, phosphorus, and potassium, respectively—of 5–10–10. Using that much wood ash will raise the pH a little, but it's at 5 now and can afford a slight boost upward. I had everything but the blood meal on hand, and decided that between the last of the Vermont 100, a couple of free sample packets of Gardens Alive! products, the high nitrogen component of the current batch of compost, and the fact that the N reading for this bed is already high, I could skip the blood meal.

I redug Ex-Boxed Bed #2—much easier this time—pulling out lots more thistle runners. There was *one* large white grub, at the edge of the bed. (Sunday I found two tiny ones in the other bed.) No worms *or* slugs; the only visible life form in the soil was the ubiquitous millipede.

I dumped "some" of the fertilizer mixture where the potatoes would be planted and raked it in. Then I spent a few minutes deciding where to put my five varieties of potatoes, each selected first for high disease resistance, and then for flavor, yield, and keeping quality, in that order.

The new *O.G. Encyclopedia* suggests a way of growing potatoes under mulch that sounds intriguing. I never saw this anywhere else, but the encyclopedia says: pile up leaves in the fall where you plan to grow potatoes; in spring plant the spuds in the partially decomposed leaves and cover with a foot of straw.

My thistle-controlling plan calls for putting that whole north end of the garden between the fence and the grass path under thick straw, which is why I've planned all along to give the space to potatoes and use the mulch method this year. But last fall, as it happens, the north end of the garden was where we put the leaves. So conditions were perfect for an experiment, and this is how I set it up:

Culture for all was the same, except that some pieces are set on bare soil and some on leaves. After sprinkling that end of the garden to give the potato pieces some moisture under their healed sides and encourage rooting to begin, I scattered almost five bales of straw over the whole planting. The plot looks like a haymow now, but a cou-

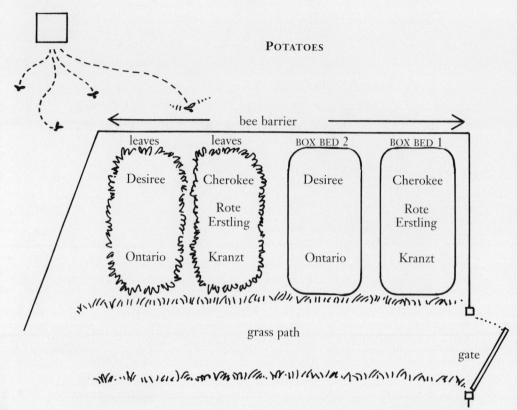

bee barrier

leaves · leaves · BOX BED 2 · BOX BED 1

Desiree · Cherokee · Desiree · Cherokee

Rote Erstling · Rote Erstling

Ontario · Kranzt · Ontario · Kranzt

grass path

gate

ple of good rains will pack it down (if we ever get any rain). I'll be really interested to see which of these methods works better, and whether the straw, which allows potato vines to grow through, will stop the thistle.

It took me five hours to do all this, but I'm happy to report that I was able to proceed in a steady, methodical way, not getting worn out, not charging at the work, not getting exasperated or frustrated (or stung!). I'm learning.

When I was bringing in the empty Vermont 100 fertilizer bag I noticed that it had an 800 number printed on it, so I decided to call the company to find out what they put in the stuff. I spoke to a guy named Steve Liffers, from whom I learned that the product is now called Vermont Organic Fertilizer, and that it's made of ground whey, cocoa meal (ground cocoa shells), something called SulPoMag, and nitrate of soda. I guess the brown powder that looked like either blood meal or dried manure must have been either whey or cocoa meal. He's sending me a free ten-pound bag—a nice return for making one free phone call.

Big news: the four Gurney Girl seeds I set to presprout a week ago have all germinated! At the other end of the seed-starting scale, some of the cabbage seedlings are about ready to start spending their days outdoors, which will toughen them up for being planted out.

April 15. Having left the bees in peace for a week, today was the day to find out if the queen had started laying again after her little vacation, if that's what it was. Liz came over for lunch and took some pictures while I opened the hive—even with the tripod and self-timer, I can't photograph myself working the bees.

There were definitely larvae present. I didn't see all that many, and they didn't appear to be in a solid pattern or of uniform size, but they were there, along with a lot of still-unhatched capped brood from last week. I *didn't* see any eggs, but Jim said the other day that he has yet to see an egg in his several years of looking; for people with poor vision, maybe that's a common problem. With so many constantly moving bees always covering the comb you're trying to look into, it must be a challenge to get a very accurate picture even for a person with perfect vision. No substitute for lengthy experience in this biz.

At the grosser level, though, I can report that several field bees were dancing on the comb—the first time I ever saw *that*, except on *Nova!* Undeterred by being lifted out of total darkness into the light, they kept right on doing what Karl von Frisch, who first described and interpreted it scientifically, calls a *tail-wagging dance:* "They run a short distance in a straight line while wagging the abdomen very rapidly from side to side; then they make a complete 360-degree turn to the left, run straight ahead once more, turn to the right, and repeat this pattern over and over."* In the language of the bees as interpreted by Frisch, the dance indicates a food source at some distance from the hive—300 meters in his experiments. Fascinating and exhilarating to see this for myself. I wonder what the food source in question could be.

Karl von Frisch, Bees: Their Vision, Chemical Senses, and Language, revised edition (Ithaca: Cornell University Press, 1971), p. 88.

Reversing was probably a good idea. All the action seems to be in the bottom box, the one that until a week ago was on top; that's where the bees were dancing and that's where I found the larvae.

Liz was working without protection—can't focus a camera wearing a veil—and a bee got interested in her right about the time I was feeling I'd upset the hive about as much as I ought to for one day. I put things back together while she beat a prudent retreat back to the house, chased but luckily not caught by a very determined guard bee.

From her own experience Liz confirms my sense that this is a pretty aggressive hive; and Liz's bee experience, though rusty, is extensive. She's my most experienced gardening friend, too, for all that she's more interested in flowers than in cabbages and squashes, and her help and expertise have bailed me out of difficulties more times than I can remember. I often wish, though, that I'd known Liz and Jeff back when they were conducting their own suburban homesteading experiment in the late '70s.

As far as I know, the term "suburban homesteading"—such an evocative one for me from the first time I heard it—was actually coined by Jeff Ball. I first became aware of Jeff's name when I picked up what is still my favorite of all his books, *The Self-Sufficient Suburban Garden,* a 1983 title from Rodale Press. (Though only Jeff's name appears on the cover, Liz collaborated on the text and took all the photographs. Their more recent books bear both their names, and a good thing too.) I realized from information in the back pages of *SSSG* that the Balls and we were virtual neighbors, and remember with some embarrassment looking up their address and driving over to sit in the street outside in my car, ogling the greenhouse I'd read about and longed to possess myself.

Alas for me, by the time Ted and I actually met them—in 1987, at one of Jeff's lectures—Jeff was already speaking of their experiment in self-sufficiency in the past tense. (Liz remembers that the first thing I said to her when I came up after the talk, in tones of thinly veiled accusation, was "How come you're not doing it anymore?") By that time they, and especially Jeff, had moved on to other projects and enthusiasms: high-visibility expertise in lawn management and residential landscape care for nongardeners, a series of instructional videos on "Yardening," a column ("Smart Yard") in *Practical Homeowner* magazine, a recurring role as the *Today Show* gardener. Meantime Liz was building a reputation in photography and garden writing. She and Jeff have turned out to be one of those rare couples able to continue a working partnership when their marriage has broken up; they've published nine books together, with more projects in the works.

The up-side of the change in directions is that if *they* were still keeping bees, *I* wouldn't have inherited all this great equipment. Actually, the stuff had already been given away to other friends, who thought they might like to keep bees someday but had never gotten around to it. But when Jim presented me with the swarm, Liz promptly called up these friends and suggested that if they didn't have any immediate plans for getting into bees, she knew somebody who did. Ted and I drove out in the pickup and returned with a truckload of beehive parts, a box of useful tools, some books (including the one by von Frisch), a pith helmet, a veil, and a hand-cranked honey extractor, all for the price of the trip.

My coverall, also Liz's, cost me lunch for two at a pleasant restaurant. Liz is five feet ten, four inches taller than I am, and this lunch-purchased garment is sized to fit its original owner. I have to roll up the sleeves and pants legs into giant cuffs. But it does the trick.

Non-bee news: one of the Gurney Girl seeds I was presprouting had seed leaves today! I poked it into one of the green Burpee cells where the Jupiter pepper seeds failed to come up, burying the root but leaving the seed leaves out, as my 1990 notes advise. The other sprouting seeds went back in the plastic bag, until they, too, develop seed leaves. As for the peppers, I'll probably buy a six-pack from a nursery. If something takes a notion to fail, just as well that it's peppers, which don't seem to need disease resistance in my garden and are easy to find at garden centers. I don't really mind about their not being organic right from the get-go.

The bee barrier is sagging and needs to be secured and tightened at the scrolled ends. Some of the magnolia blossoms are fully open and look beautiful; others still have the little fuzzy cap on, holding them shut. There's a storm coming; hope the open blossoms don't freeze (they seem to at least one year out of two, this far north). For once I'm happy that rain's forecast—the onions and too-fluffy straw and asparagus all need water. Two asparagus crowns in the row along the fence have sent up shoots that will certainly be edible in a couple of days. Can the others be far behind, especially if we get a decent rain?

I've realized something surprising, working this spring. I'm not doing the Homestead Year out of devotion to a particular piece of land. I don't expect to remain on this land, any more than I expect to remain on the earth. I wish there were a spot of land I could belong to in the way my role models do or did: Wendell Berry's farm and Harlan and Anna Hubbard's homestead in Kentucky; in New England, Noel Perrin's South Farm, Donald Hall's grandparents' farm, Scott and Helen Nearing in Vermont and Maine. I admire and envy that, but it isn't for me. I don't believe in permanence. But I believe in the activities and arts of this life, and here is a fine place to practice them. For a while.

April 16. Rain still expected, I hope for good reason. This morning, I went out under the leaden sky and reinforced the bee barrier.

An addendum to the Gurney Girl saga. In future, if I wish to grow this variety, I would be well advised to eat my words about starting seedlings in flats of soilless gro-mix or whatever and potting them up in soil as they emerge. Since I'm having to presprout these seeds with heat in order to germinate them anyway, which also means having to plant them once the seed leaves emerge, it would have been *much* less trouble to have begun with flats in both cases. Ditto peppers, I now think. The other crops, even the eggplants, do well enough by my present one-step method, but Gurney Girls and all peppers should henceforth be coddled if I want to save myself a lot of bother. REMEMBER THIS.

One tiny speck of green in one white pot of potting soil: pennyroyal.

April 17. Rain in the rain gauge this afternoon: 0.15 inch. Almost too little to care about, if it weren't the first measurable rain this month. April showers, ha! Yet for that reason or some other, the onions have put on a growth spurt.

Both the pin oak and white oak are suddenly flowering, and the bees appear to be harvesting their pollen; field bees are coming into the hive with pollen baskets packed, some yellow, some white. The hive appears to be totally uninterested in the pail of water with a board in it, that I left for them because of the drought. Three days ago, when I rigged the hose to wet down the potato planting area, I saw bees drinking from the leaky coupling at the patio faucet, but they don't appear to like the bucket.

The cabbage seedlings are mostly ready to harden off. I sorted out those big enough (15) from those that need more time under lights (5) and brought the graduating class up from the basement; tomorrow, rain or shine, out they go.

April 18. Another 0.15 inch of overnight rain in the gauge this morning. I checked when I took the garbage out to the compost bin, and picked the first asparagus spear about fifteen feet from the beehive on my way back in. The bees were sticking close to home, though I saw one crawling on the violet-strewn grass, perhaps drinking the cold rainwater. I stuck one bare hand under the outer cover to fetch out the empty syrup jar, and a bee flew out and landed on my wrist. It was all I could do not to jerk away and run, but I didn't, and she didn't bite. I set the jar down in the violets to give the hangers-on a chance to leave. Tomorrow's supposed to be warmer; if it is, I'll dose the bees again with Terramycin and give them the last of the syrup laced with Fumadil. This medicating needs to be finished four weeks before the first honey supers go on, so no antibiotic-laced honey winds up in the kitchen.

The Red Sails and Black-Seeded Simpson lettuce seedlings have true leaves and were ready to thin. I'd sown the seed liberally because it was all several years old, but it came up liberally too and I had to sacrifice a lot of healthy, pretty little plants to get one sturdy seedling to a pot. Always hard, but it's done now.

The cabbages went out into the raw, cloudy day for their first view of the big world, then back in at dusk.

April 19. Today, for the second day in a row, The Weather Channel promised it would warm up and clear up in the afternoon. Instead we got another high of 50 with a wind-chill factor making it feel like 37 out there. Phooey. The jar of syrup and bowl of antibiotic-laced powdered sugar are sitting side by side waiting to be taken to the bees, but even gentle bees would be grumpy on a raw day like this, and I know better than to open this hive, even just the top.

I started five pots of Janie marigolds in potting soil, putting them in some little square plastic nursery pots I found in the basement and scrubbed up yesterday. When it's warmer I'll direct-seed some more of them, but it'll be nice to give these five a bit of a head start. They're on my study floor, enjoying the bottom heat. I'd have started more now if I'd had more pots on hand. This is 1988 seed, so I put four or five seeds in each container in hopes of getting one to grow.

This morning, 0.3 inches in the rain gauge. Total for the past three days: 0.6. Good! The onions continue to respond.

Last year the blueberries were blooming by now; this year we're only at the loose-bud stage. The Prima and Liberty apple trees give signs of flowering, but the Golden Delicious tree, alas, has very few flower buds. The three or four apples it produced last summer *were* so delicious that I was hoping it would do well this year. Maybe it's just slow.

I'll be planting cabbages outside within the week, meaning we must deal with the woodchuck threat *now*, either by putting down another flat length of one-inch chicken wire they can't dig through, or ordering electric fencing. Tomorrow I'll call and get an estimate. Today I went out and paced off the perimeter: 79 long strides, or nearly 240 feet . . . call it 250 to be on the safe side.

April 20. I spent half an hour on the phone this morning with a sales-
man at Premier Fence Systems, working out a "system" for the garden
that will deal with woodchucks—and, incidentally, any raccoons that
get interested in the sweet corn, and maybe even with the Purbricks'
cat, who thinks of the garden as her own great big litter box away from
home. A single strand of charged wire six inches above the ground will
do the job, when placed outside an existing fence. And a solar panel actually turns out to be *cheaper* than an ordinary charger. The "pain potential" is much greater with a conventional system, but this should do
the trick; I only want to discourage woodchucks, not exact revenge. And
it's a chance to get into solar power in a tiny way.

 The cost is going to run above two hundred dollars, but chicken
wire costs something, too, and would be much, much, much more laborious to install, even just flat on the ground. And we'd have to keep
a two-foot-wide strip of dense mulch around the whole fence all summer because the mower wouldn't be able to get any closer than that. To
avoid that necessity, we *could* do what the *O.G. Encyclopedia* recommends
for stopping woodchucks from burrowing in: dig a trench a foot deep
and two feet wide around the whole 250 feet of perimeter! You exca-
vate the trench, bury the chicken wire, and put the dirt back in. I don't
think so. How long before buried chicken wire rusts out? And we
thought digging the fishpond would be a big deal!

 All the alternatives seem less certain (and less portable) than
the electric fence. I'm inclined to spend the money, if Ted will go along.

 Another raw day that refused to clear up as forecast. It did get
warmer toward late afternoon. The books all say to open a beehive between 10:00 and 2:00 on a bright sunny day; but I thought about it, then
suited up and went out there around four, with only my hive tool and
the little dish of Terramycin and sugar. No smoker. The bees didn't like
being disturbed—they never do—but I moved very slowly: took off the
outer cover and the empty hive body, removed the feeder jar with adhering bees, pried up the inner cover (making a number of them mad),
and very gently sprinkled powdered sugar on the ends of the top bars

of the frames. They got interested in the sugar right away. I then carefully put things back together, picked up the jar, and retreated. No problems, and the antibiotic that works against foulbrood is finally in the hive.

More pennyroyals are up, and tinier sprouts I never saw.

The tomatoes in the Burpee starter need to be transplanted into bigger pots—something else I should have foreseen. Time to pitch that little green twelve-pack; it's done its time. I rummaged around in the garage and brought in an assortment of small plastic and terra-cotta pots to clean up over the next day or two.

April 22. The bees have discovered the old hive bodies, filled with wax-moth-riddled frames, that once belonged to Liz and Jeff. They've been stored in a stack all winter next to the garage, and riddled or not they're full of wax and propolis, and smell like bees to bees. How convenient if the hive would swarm and take up residence in there! I wouldn't have to lift a finger or lose a bee. (But how inconvenient to have them whizzing back and forth right across the main part of the backyard.)

The marigolds are up. Two more asparagus spears came in, with more due tomorrow. One more stalk has broken the surface, but it looks diseased; that crown may have to come out.

I have a rotten cold and stayed in all day, though this afternoon was hot and muggy after a dark, wet morning (0.3 inches of rain overnight—that's almost an inch since Friday). Since I wasn't able to work outside, I spent the time making a rough and tentative plot of this year's garden and a planting schedule. I also called Premier and ordered the electric fencing plus solar panel.

Two more cabbage seedlings made the grade and came up from the basement.

The soil test results are back from Penn State. To my surprise, things look pretty good. The pH is 6.7—much higher in the garden proper than in the potato patch. All they're recommending for fertilizer is a pound-and-a-half per one hundred square feet of 10–10–10. Each big bed is about half that size, the little beds somewhat less. They don't give organic equivalents by weight but I guess I can calculate them with all the reference books I have at hand. Evidently the potato patch is in poorer shape generally. The boxed beds were made in 1989 and limed that year and not since, which could explain the lower pH now. I took the soil samples for both my own and the Penn State tests very carefully according to instructions, so they should be as accurate as these things ever are. The potato-patch area has been cultivated much less, meaning I've added less stuff like wood ashes and bonemeal to it—maybe that accounts for the difference in mineral content. Something must.

April 23. Another varied and productive day on the old homestead. It was a beautiful spring day, blue sky and sunshine, trees and shrubs from apples to azaleas right on the brink of blossoming. I started with a tour around the yard, looking with a critical eye for whatever needs to be done. Helen Stipe, our elderly next-door neighbor, saw me and called me over to say how much she'd enjoyed the magnolias this year—they weren't frost-killed despite our fears and are now just past their peak, petals thickening on the ground. Then she asked how the bees were doing. I was a little worried that she was going to raise objections, but not at all, she was interested: "It's a whole different world, isn't it?"

To my great regret, the Prima tree, source of last year's good crop, has almost no flowers coming this year. The Liberty has lots of flower buds, but is a much smaller tree. None of the others appears to be going to flower either. So it isn't the late pruning. It may be the *severity* of the pruning, or—the likeliest thing—a combination of overfeeding and severe pruning that stimulates vegetative growth at the expense of fruiting. Whatever, I'm terribly disappointed. Last year's applesauce was such a success. This year we mustn't feed these trees at all. Ted will have to think of something else to do with the grass clippings.

Like put them on the blueberry hedge. These mostly unfed bushes don't grow much, though they make lots of blueberries. The bushes nearer the red maple have far fewer buds than those at the other end of the row, which probably gets a lot more sun. I got the grass clippers from the garage and trimmed the invasive grass back at that sunnier end—deeper mulch needed there.

I scrubbed out the dirty flowerpots that had been soaking overnight in Spic and Span, to kill any malingering disease organisms from previous years, and carried them out to dry in the sunshine.

It's two weeks today since I opened the hive and reversed the boxes. My second chore was supposed to be slipping a queen excluder in between the two hive bodies, preparatory to splitting the hive in two. All swarm-control systems require locating the queen, and a queen excluder is supposed to make life easier for beekeepers like me, who have a hard time finding the queen. The excluder looks like a little grille, with bars spaced widely enough that worker bees can move between them, but not queens and drones. It confines the queen to whichever box she's in when you introduce it; all you have to do is wait a few days, then open the hive and look for eggs. The hive body that has eggs in it is the one where the queen is.

I reviewed my books and notes and took my brand-new queen excluder out to the hive before recognizing the fatal flaw in this management approach: I've never seen an egg, any more than I've ever seen the queen. With eyes as bad as mine, it's not easy to see even the very young brood, underneath all the bees. You're supposed to brush them gently off the comb with a bee brush (but they don't stay brushed), and to stand so that the sun's light shines over your shoulder, directly into the cells (but I never seem to think of that till it's all over).

In the end I tried to combine two different approaches I'd read about, but it's plainly a case of too many experts counseling the ignorant. The hive's standing out there now, *three* hive bodies tall, with a super (filling with honey, probably from the sugar syrup) on top of the stack. I still need to divide it into two discrete hives, but am paralyzed with indecision. How? Which way?

The trouble with waiting to choose a method is that the tiny uncapped brood I put in the top hive body—something both experts said to do—will quickly get too big to be fed royal jelly and raised by the workers as queens. The jelly diet has to begin within the first few days of hatching. I'll have to go in again and find some more day-old brood if I wait, but I can't decide how to proceed.

Exhausted as usual after dealing with the beehive, I was ready for a tamer chore: digging the beds I'll need next for planting carrots and garlic. And for transplanting cabbages, which look healthy and beautiful and are really ready anytime to go out into the garden, but have no place to be put and no electric fence as yet to protect them from woodchucks. I consulted my plot of the various beds, then got rake and spade and sallied forth to dig up one bed and another half-bed in the main garden. Both of these, like the rest, were shaggy with beautiful deep-green hairy vetch, the nitrogen-rich, cover crop of "green manure" I sowed last summer, and also with tall flowering weeds. I dug vetch and weeds under indiscriminately (except for the dandelions, which are providing a good honeyflow this week). I cut the vetch first with the scythe—not very efficient, the scythe is much too long-handled for me, especially when working raised beds—then had to jump vigorously on the shovel to make it cut through all that freshly hacked-off vegetation into soil. Something like mining peat; often I had to make three cuts on three sides in order to lift and turn one spadeful of dirt and plants. The soil was actually too wet to dig—it would have been better to wait a day, but more rain is predicted for tomorrow, and the job needed to get done.

With the two beds of vetch dug under, I turned to the next chore: making potting soil to use for transplanting the tomatoes. Not a fussy job. I half-filled a bucket with soil from an unturned bed and added several double handfuls of nutritious garbage-derived compost from the supply in the storage cans, and several more of much less nutritious wood compost (mined from under an old woodpile) as a lightener. Then I fetched forth my root-bound Beefmaster tomato seedlings and potted them up with great dispatch in nice roomy pots, first putting little pieces of newspaper in the bottoms to cover the drainage holes.

The little Beefmasters spent the rest of the afternoon on the sunny patio before being returned to the lights in the basement. In another two weeks it'll be their turn to be hardened off, but right now the cabbages and lettuce—which I brought up on the return trip—are enough to handle. The leaves of the tomato seedlings, all varieties, are turning purple underneath, a nutrient-deficiency condition I remember from every year I've started seeds. They get over it once they're planted out. And the Beefmasters, in hog heaven with all that new compost, will start getting over it now—their new leaves will come in green.

In other years I've had tomatoes in the ground by May 1, but this cold, peculiar year it seems better anyway to wait till the official last frost date, May 15. The largest lettuce leaves are good-sized, six inches long or so, but very tender; they usually wilt and die from sun and wind during the hardening-off process, but the roots aren't affected, and the new leaves come in tougher. The next thing to dig will be the lettuce bed.

My final project for the day was the messy but satisfying one of cleaning up some of those stored wooden frames. Their comb and foundation is rotten from wax moth damage; I wanted to remove the long side bars and the mess—damaged comb, cocoons, webs, and frass (droppings)—and scrape the top and bottom bars clean of wax and hardened propolis, in preparation for fitting these with new, short side-bars and new sheets of foundation. Then they'll be the right size for the honey supers I'm going to build.

It's interesting, agreeable work, and one feels pleasantly thrifty, recycling the woodenware instead of junking it. I brought a chair outside and worked on the patio for a while, knocking the frames apart with a tack hammer, pulling the damaged comb loose, scraping away at the lumps of propolis with my hive tool. When I saw what a mess I was making, I moved the chair into the grass, where bits of wax and propolis, and frass from the wax-moth larvae, wouldn't land on the flagstones and get tracked into the house before I could sweep them up. Propolis is awful stuff to work with. The hive tool and hammer are sticky all over with little amber flakes. The jeans I was wearing were covered with them, and went straight into the laundry basket when I took my Lyme-disease-preventing (and propolis-removing) shower before dinner.

I didn't get a lot of frames finished, but expect this project will prove to be like the many others—paper-grading, for instance—where getting started is the hardest part.

April 24. This morning, after putting out the cabbages and lettuces (like so many little dogs and cats) for the day, I opened some cartons of stuff ordered long ago from the Kelley Company and assembled the frame-nailing device, a fiendishly clever little box that holds the end bars of the frames while you nail them together. Very conservative beekeepers would use glue as well as nails when building frames; but if Jeff had glued these frames together, I wouldn't be able to get them apart now without breaking them, so I'm sticking to nails only. This soft, clean, beautifully milled wood is a pleasure to work with, and it's fun to unite the new wood with the scraped-down old wood, and nail the pieces together with my little tack hammer, the perfect tool for the job.

I nailed all the pieces I'd cleaned up and was set to clean some more, but had errands to do later on and decided against getting gucked up with propolis beforehand. So instead I straightened out some pieces of plastic-covered wire that had spent the past year twisted around the top strand of the garden fence, using two pairs of pliers to pull them straight, and the wire cutters to make lengths suitable for fence repair. I might have then gone out to make a start on this necessary project, but wanted to take a walk and mail some letters before the forecast storm arrived. (It never did.)

Returning along the trolley path, I heard a crashing in the leaves and saw a woodchuck—he looked big as a beaver!—running like crazy through the woods and up the hill. At nearly a mile from the house it was probably not "our" woodchuck, but I don't really know how far they range. The one that plagued us two years ago seemed to live right next to us. I met him once in broad daylight, strolling through the cemetery behind the house; he stopped for maybe half a minute—I stopped, too—gave me a surly glare, bold as brass, then broke eye contact and sauntered on at his own pace.

April 25. Rain 0.1 overnight. This morning I gritted my teeth and prepared to assemble the shallow supers that came in the shipment with the nailing device, way back in January. I ordered them because I want to switch over completely, and use shallow-depth supers instead of full-depth hive bodies for the whole hive, and not just for surplus honey. It's a practice not uncommon among women, older men, or anybody with a bad back, and the bees don't seem to mind. A hive body full of honey and bees weighs around sixty pounds; a full super weighs more like forty.

The pieces of wood look like pine: lightweight, clean, fragrant, very soft and easy to nail (or beat up with a misplaced hammer blow). They didn't come with instructions, so I got out *Keeping Bees* by John Vivien—with Gene Logsdon, one of my two favorite homesteading writers—and followed the instructions given there. First you assemble all your tools: big hammer, tack hammer, bottle of wood glue, nails of various sizes, hive tool, some kind of square, and block of wood to hammer on so you don't scar the soft pine. Then you put glue on the insides of the finger joints that join the four pieces and knock them into a tight fit. This is simplicity itself, except for how easy it is to put things together so that one of the cut-out handholds or frame rests winds up upside down.

Then you insert nails of the correct sizes into the predrilled holes of the finger joints and pound them home, checking constantly to see that the box is still square. This part ain't so easy; the first one I built is off square by a degree or two. And the last step is to nail the nine-frame spacers to the frame rests.

After dinner, while Ted went shopping, I assembled another super. Only then did I realize how much easier it is to keep the pieces square if you stand the box on end and nail the nails straight down, rather then pounding them in from the sides. This must be such basic carpentry that even John Vivien, who does everything for you including draw you a picture, forgot to mention it. Anyway, my second super is beautifully square, and I found I was getting good at whacking the nails into the wood.

In between assembling the two supers I redug the cabbage bed–to-be, incorporating numerous buckets of compost and raking the whole thing smooth. The bed's ready, the cabbages certainly are; what's not ready is the fence, and what's needed is a nice dry day on which to work my way around the whole perimeter, wiring the Bunny Barrier tight to the fence proper.

Our garden fence is made in layers, each layer added as we discovered a hitherto-unanticipated problem in the existing structure. When we first made that garden, five years ago, we enclosed it in serious metal fencing: heavy-gauge wire woven in meshes two inches by four. The fence stands three feet high and is supported by those green metal stakes that have flanges at the end you pound into the ground. It was the sturdiest garden fence we'd ever built, and it expressed our sense of the seriousness and permanence of this garden compared to previous ones.

It wasn't very long until we began bumping into rabbits in the lettuce patch. How were they getting in? We paced round and round the fence, looking for holes, gaps, places where they might have dug underneath. Only after long bewilderment did we actually see a rabbit, panicked by being caught in the garden with us, dart straight *through* the two-by-four space in the wire mesh without even breaking stride.

Rabbits never went through the makeshift chicken-wire fences at our previous house. Chastened, we went out and bought a roll of four-foot hex wire of the kind we'd used before. We cut the whole hundred feet down the middle and wired the chicken wire to the bottom two feet of the existing fence, all the way around. A tiresome business, hard on the knees, back, and fingers, but we'd learned our lesson.

For a time, all was well. Then one day, behold! a bunch of broccoli plants nibbled in a familiar fashion, and a rabbit kiting around in terror inside the garden. About the same time I read, in an article about rabbits and fences, that meshes larger than one inch are too big to keep rabbits out if they really want to get in. The roll of chicken wire we'd bought had two-inch meshes.

Back to the hardware store we went, and home we came with

two hundred feet of one-inch hex wire one foot tall. And back around the fence we crawled, wiring the chicken wire to the chicken wire to the mesh fencing and binding in a lot of weed stalks in the process. It's this last layer, fastened in haste and aggravation, and mostly with twist'ems of paper-covered wire, that has come loose. Before I put lettuce and cabbage seedlings into the ground, I need to do what's necessary to ensure they'll still be there the morning after.

That's not an easy thing to ensure, even with three layers of fencing at rabbit height. For one thing, woodchucks dig *under* fences. For another, the very tiniest baby rabbits can squeeze through even one-inch hex wire, and do, a couple of times at least, in the week or so before they get too big. The passive solution would be to fence the garden with hardware cloth—if we'd done that in the first place, it probably wouldn't have cost any more than the three layers of fencing added together—but I'm willing to stop where I am as far as rabbits are concerned.

Woodchucks are another matter. We've tangled with *them* before, too. I guess I'm going to gamble that before a woodchuck—the one we saw a while back or another one—discovers this year's garden, the electric fence will be installed.

After digging the cabbage bed, I also dug compost into the non-trellis half of the trellis bed, and dug the vetch under in Compost Bed One, where popcorn is to go. Then I managed to get Ted to stop grading term papers long enough to mow the grass of the turf paths, as well as scythe the vetch in the future lettuce bed—the scythe is sized for him, not me. When he'd finished, I turned the vetch under, or at least did my best to incorporate those long green strands into the soil. As usual not so much as a white grub was stirring underground.

April 26. Nearly 0.5 inch of rain in the gauge this morning, and the onions have really taken off. After breakfast I assembled one more super (not perfectly square, but almost) and cleaned up my work space on the dining-room table. Two more supers to go; then I'll paint the lot of 'em.

April 28. Less than 0.1 inch of rain overnight. The cabbage seedlings are bigger than I *ever* remember letting them get before planting them out. Today was the day of rabbit-proofing the fence. I got out my bag full of pieces of straightened plastic-coated wire and worked my way around the entire periphery, clearing leaves out of the gap between fence and chicken wire and wiring all the layers tightly together. While I was at it, I trimmed back the honeysuckle that covers the stretch of fence by the strawberry patch and the former Island; it will have to be watched all summer to be sure it doesn't interfere with the electric wire. I hated to do that before the vines bloom, but there wasn't any choice.

That took up the morning. This afternoon, after Ted mowed all the way around the fence line, I mulched the foot of ground outside the fence with last fall's leaves. Back and forth I went, loading up the garden cart with wet leaves, trundling out to dump them along the fence line, loading up again.

The bees, intensely stimulated by the sunshine as always, were positively *fountaining* up from the hive entrance; if you stand in the right spot you can see the light on their wings as they hurl themselves out and up and away. Impressive. Thrilling. Frightening, even, they're so obsessed and fanatical when the sun's strong and there's a flow on.

Before starting on the fence I checked the super—nearly full of honey, and heavy as a couple of sadirons! This is one hell of a strong colony. I *must* reach a decision and divide it between now and May 12, since I'll be post-op and out of action for a while after that. I'll have to try the only way I can think of that might work: sort through all the boxes and create two fairly-well-balanced colonies, putting very young brood in both (or a queen cell, if I find any). That way, whichever colony the queen doesn't end up in can raise its own new queen. The queen-less one will lose a little time, but it's better than letting them swarm.

Putting the old queen in one hive and giving the other hive a new queen from a breeder would probably—no, certainly—be preferable; but without the help of somebody with sharper eyes than mine, I'm just not going to be able to manage that. A viable hive has to have a queen, but won't tolerate more than one. If I put the new queen in with the old one by mistake, one will kill the other, and we're back to square one. If I bought *two* new queens and put one in each hive, maybe I'd end up with two viable hives . . . but that's as good as an admission of defeat, and I guess I'm not quite ready to throw in the towel.

There's no indication that swarming is imminent. They're working their little butts off, not moping or clustering around the entrance. I think we're okay for the moment, but between now and May 12 I *must* proceed as described.

Other news: some of the blueberries, the kind with white flowers, have bloomed. (I keep meaning to identify the varieties, since some of the tags have vanished and I don't seem to have a record of which five types I ordered. Nor does the invoice specify types. Northblue, Blueray, Bluecrop . . . and two more. If those are even right.) Buds are forming on one of the currant bushes, the hardy kiwis are greening up like anything, and the Liberty tree is going to be in full bloom in a couple more warm days.

The strawberries are also beginning to flower. Not many plants still alive this spring; the leaf spot was pretty bad last summer. This is their last hurrah; when they're done this time, we dig up the patch and plant melons and cukes. Tomorrow the cabbages go in the ground; lettuce, too, if time allows. Then carrots, sweet peas, dill.

The electric fencing arrived today, though I haven't yet opened the box. Heaven grant that the woodchuck (our next-door neighbor, Jan Purbrick, has seen him, too) won't discover the greens till we can make time to rig the fence.

April 29. A stunningly beautiful day, perfect weather for our visitor from Sweden. Lars-Håkan is a poet and an instructor in the English Department at the University of Lund, where I taught for a couple of years myself, back in the Sixties. He's in the States for a conference at Kalamazoo (which Ted, incidentally, will also attend). Lars-Håkan is interested in poetry almost to the exclusion of all else—his Kalamazoo paper is on Spenser—but this afternoon he good-naturedly helped me plant out the cabbages and lettuces: sixteen of the former, eight of the latter.

I've done this so many times the procedure goes quickly and smoothly. Scoop out a hole with the trowel and dump in a double handful of compost and a handful of Square-Foot Fertilizer. Stir these into the loose soil in the bottom of the hole, pour in a bowl of water, and, while the water's sinking in, tear off the part of the peat pot that sticks up above the potting soil (so it won't wick moisture away from the roots). Pinch off the seed leaves and the lowest two true leaves and wrap the stem in a strip of newspaper to form a cutworm collar. Set the pot in the glop at the bottom of the hole and refill, pushing soil around the stem so that the paper collar is half above- and half below-ground and the soil forms a saucer around the little plant. Usually I then water again, but this time the soil was so wet I didn't bother. And so to the next.

Willing assistant, Lars-Håkan poured water, tore off pot tops and handed the pots to me. After planting six cabbages I got called away

IF POT PLANTED SO:

peat draws moisture out of soil
into air, roots dry out.

H_2O

soil line

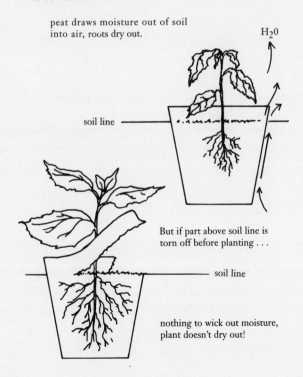

But if part above soil line is
torn off before planting . . .

soil line

nothing to wick out moisture,
plant doesn't dry out!

to the phone. He'd been watching closely, and while I was gone boldly planted a cabbage all by himself, "by the book" as he said. A good job, apart from the collar being a bit low to foil cutworms. Later he planted a couple more while I took another call. I asked if he'd like to do the last one, and winced as he tore—not pinched—the leaves off and wrapped the newspaper collar around with a less-than-sensitive touch; but this is just a mother's natural concern for her infants. They'll be fine.

I wouldn't let him do any lettuces, they were so delicate. They were also in plastic pots filled with Jiffy mix, which tends to fall apart despite the network of roots. Vetch had been too recently dug into the bed and made a dense mat of unrotted vegetation beneath the soil surface. I had to stab the trowel through the mat, and the resulting planting holes were not very big; nor was the clay soil well mixed with compost—mixing was impossible. I worry a bit about air holes, but I planted them anyway. Sink or swim. They should do okay, too.

At each node on the stems and branches of the hardy kiwis there's an inch-long leaf cluster. Got to trim off the ones on the stems, which should rise smoothly to the top of the pergola.

Under their snood of dead vines the lilacs are blooming.

April 30. This morning I finally worked out the steps I'll have to go through in order to make the divide. I'm starting with a single hive, consisting of three hive bodies plus a super almost filled with honey, some of this honey undoubtedly made from medicated sugar syrup—they were fed so much of it so recently that some was bound to be stored. I want to end up with two hives: one with two hive bodies, another with a hive body and two supers (including this one with the medicated honey). I'd have tackled the job this afternoon, but the day never warmed up enough—the brood will be exposed a long time, it'll have to be at least 70° and sunny, with very light winds. Warmer would be better. Now I'm off to New York for a literary weekend, but I don't *think* they'll swarm before I get back. Maybe next Monday, May 4.

In the rain gauge, 0.2 inches. I peeled back the straw to see what was going on in Potatoland. New sprouts rising from the cut pieces, not much progress for sixteen days but of course the deep mulch keeps the soil from warming up.

I planted the elephant garlic, thirteen cloves from the three expensive ones we set out in the fall of 1990. (One died of a neck ailment; the other two got enormous.) I spaced them six inches apart and stuck them unceremoniously into the end of the cabbage bed. Then I redug six feet or thereabouts of the trellis bed, planning to plant carrots there. To that end I dumped on a lot of wood compost, along with the rest of the bucket of already-mixed fertilizer, before turning it over. The soil was so clogged with vegetable matter—though not nearly as bad as where the lettuces went in—that I decided to plant only Thumbelina carrots, a round type that won't have to penetrate deeply—I got freebie packets from both Gurney and Jung, Thumbelina is Trendy Gourmet Variety of the Year—and wait to plant the long Nantes types for fall, when the soil will be looser and I can incorporate sand. I made three parallel shallow five-foot furrows with the trowel, doled out the seed (the whole Gurney packet and part of one from Jung), covered it with a quarter-inch of wood compost, and covered the bed with boards, to keep the soil moist and keep the seed from washing out till it sprouts.

Lastly I drove two six-foot stakes into the end of two beds and planted sweet peas around them at three inches, one inch deep, just like pole beans.

I grubbed around in the dump next to the Purbricks' and found another cinder block. That one plus the one in the fire pit, which was used to support the sap pan, will be the base of the new beehive. I carted the two of them over and stashed them by the corner of the fence to await the great day.

For dinner this evening, the first asparagus. We finally collected enough for two respectable servings.

$\mathcal{M}$ay

May 3. This afternoon I returned from three days in New York—three hot, sunny days, *perfect* for making the divide, as I kept thinking while walking across Central Park and around in the Museum of Natural History, accompanied by various urban poets who know zilch about bees and care less—and plunged into gardening almost as soon as I could change my clothes.

Here's what got accomplished: after bringing the tomato seedlings up from the basement, all but the still-tiny Gurney Girls, I made compost tea in a bucket to medicate the twelve I wasn't going to transplant right away. All have very purple leaf undersides, indicative of that nutritional deficiency. If you soak a couple of handfuls of compost in water, a lot of its nutrients leach out and the water's a great tonic for puny plants. (Nutrients in the soil will pep up the ones going straight into the garden.)

Outside, Ted scythed all the remaining vetch—not a great way of mowing vetch, incidentally; too many strands lie too flat to the ground for the blade to catch it all. He worked his way up one side of the beds and back the other, but some strands still got missed. (He looks very graceful scything, I must say—like a dancer.) When he was done I piled all the cut vetch from two adjoining beds onto Bed Five, a super dose of green manure. Then I turned another bed under, working all the while within the cloud of heavenly fragrance exuded by the lilacs.

Ted had mowed the grass while I was away and mulched the cabbages, onions, and lettuces lightly with clippings.

Those leaves the kiwis were putting out at nodes along their stems have become tendrils and must be pruned off before the vines expend any more energy growing them. Blooming: the Liberty, the earliest azaleas, the strawberries, and the blueberries, all varieties. Tonight we enjoyed our second asparagus-enhanced dinner.

May 4. No bee day this: cool and overcast, high somewhere in the low sixties. Bees sulky and not foraging very well. Phooey. I asked Jim last night if my strategy for making the divide sounds okay, and he said it did. Also that the middle of May to the middle of June is the main honeyflow season around here, i.e. the period when the most nectar-producing plants are in bloom, but that the honeyflow has begun already. A good honeyflow stimulates the swarming impulse. *I have got to make that divide!*

The purple-leafed tomatoes went outside this morning. I strained the compost tea and watered them. They may get planted outside without much hardening off, depending on the weather and the week's work. I did plant out the last four cabbages, bringing the total to twenty and nicely filling out the bed.

My principal project for the morning was to dig another growing bed in the main garden, one of the ones Ted scythed yesterday. *Very hard work!* It took a long time, and I'd had a short night, so got extremely pooped. Hairy vetch is a legume, which is why it's valued as a cover crop: legumes "fix" atmospheric nitrogen in the soil by means of little nodules on their roots; it's free fertilizer. These roots form a tangled mat that holds the soil, and that soil-holding tangle is the other reason cover crops are so great over the winter—they stop erosion—but also why turning them under with a spade is difficult. The soil doesn't fall handily into the hole; you have to invert the whole large clump of soil and dump it, and then it won't mix up into a nice blend of clay and compost. This will all have to be redug when the vetch (and weeds) have rotted down.

This afternoon I mulched one path with newspapers and straw, after first digging up the dandelions and thistles. Even Canadian thistle shouldn't come through the paper. After that I dug another bed, which went better than the first. Then I called it a day.

The garden's starting to look tended: two beds planted up, most of the others turned, all beds mowed. When I get the tomatoes in and the cages on, mulch the other paths, and rig the electric fence, then it'll be ready to be left alone while I'm convalescing.

N.B.: This hard labor is joyless on too little sleep.

May 5. One more growing bed dug, as well as the triangular plot that wrecked my elbow two Junes back. I hadn't planned on planting that space, didn't fit it into my rough sketch of the garden, but when I'd worked my way back that far, I thought: why not? Last summer, when I dug all the beds to sow the vetch, I stopped before digging the backmost one and the triangle because my tennis elbow was howling so loudly and the doctor said I was undoing the good the pills were trying to do by continuing to irritate it. So that back corner of the garden between the fourth bed and the compost bins was all grown up in blackberry canes, which had woven themselves right through two of the tomato cages. I got pruners, loppers, and hedge clippers—also leather gloves—and chopped through all that briary tangle to free the cages. Then I fetched forth my secret weapon—Ted, who brought in the little mower and mowed the whole mess into fine particles I was able to dig right into the soil. Sometimes a machine is just what's wanted, if it's old and beat up and you already own it.

I also pruned the rose of Sharon flat to the fence. It's grown up through the meshes on both sides of the garden. I like it there— our own modest hedgerow—but not sticking out in my way. Lots of thistle runners in both beds, but I knew that already and have planned on covering that whole back part of the garden with black plastic, beds, paths, and everything, once the vegetation has had time to break down so I can dig in some compost. After my operation, that means— around June 1.

Asparagus for dinner again tonight.

May 6. MADE THE DIVIDE!!

Strictly speaking, the day wasn't warm enough; but the sun was very strong and I'd guess that the air temperature over by the hive hit

the high sixties in mid-afternoon. With rotten weather forecast for almost the entire time between now and next Tuesday, I decided it was
now or never and took the plunge—drove Ted to the airport a bit early,
in fact—he's off to Kalamazoo—so's to have time to spring into action
if I decided conditions were sufficiently favorable.

I'd wanted an accomplice to read my instructions to me, but
there was none available so I just clipped the pages to my clipboard and
brought it along, and consulted it between steps.

Again, I used no smoke. I did everything according to my plan,
down to the point where I was supposed to divide capped and uncapped
brood evenly between two boxes. Then the usual problem arose: how
to *see* what's on the frames underneath a solid *moving* coating of bees? I
did find some uncapped brood to put in each hive—none very, very
tiny, though—and lots and lots of honey. And I kept thinking I'd finally
found the queen, but it always turned out to be a drone. Fair number
of *them* walking around on the comb.

I don't know whether any of the larvae are young enough to be
turned into queens. I do know I found two queen cells—real ones this
time, and apparently with something in them—on adjacent frames, and
tucked them back into the parent hive. Only after I'd got everything
back together did it occur to me that I should have put one of those
queen cells in the new hive and one in the old. Idiot! Jim concurs that
that would have been a good plan, one I can still put into effect this
weekend, when I go back in to have a look.

(While making the divide I noticed that the bees had a hornet
on the ground and were fighting it by chewing off its legs. Afterward
there was a legless dead hornet and a little pile of dead bees in front of
the hive.)

It's going to get cold tonight, possibly there'll even be a frost. I
hope the brood makes it through all this rough treatment.

Before and after doing all this, I mulched two more paths with
newspapers and leaves from the leaf mine. I don't much like using newspaper—brittle bits of it from last year's mulching are all over the garden—but it works well and biodegrades dependably, and we've got

major problems with dandelions, wild onion, violets, thistle, and nut sedge—all stubborn and invasive, all able to break right through a layer of leaves or straw. So I'm bowing to necessity, but I still don't like it.

May 9. Yesterday was a cold, raw, *very* rainy day, passed productively in clearing my heaped desk of piles of paperwork—work I'd been putting off for just such a day. In the evening I finished gluing and nailing the last two supers together—one doesn't sit flat for some reason, probably one of the pieces of wood is a little warped, but a full load of honey and bees should flatten it out nicely. I forgot to check the rain gauge, but will do so in the morning.

Crop report first. After that all-day rain, slugs have found the cabbages, which are much chewed. I checked all twenty anxiously to see if the growing tips of any had been eaten out—this has happened in the past with young cabbages and lettuces—but all appear to have survived. The lettuces aren't putting on a lot of growth—too little sunshine, too little rain up to now—but the new leaves are much denser and more crinkled than the wimpy ones grown under lights. So far, no losses to cutworms.

I finally took the pruners and spent maybe three or four minutes snipping off the tendrils that had emerged from the stem part of the kiwi vines. Kiwis are supposed to devote all their energy to top growth, and now they can; last year when I did this, the stems remained bare for the rest of the season. I look at the apple trees, four of them with such pitiful handfuls of flowers, and regret terribly that I didn't leave at least one of the others—the Liberty or the Golden Delicious—unpruned besides the Prima.

The first potatoes—a few Krantz and a few Rote Erstling—have burst through the straw in the form of beautiful deep-green leaf rosettes, one of my favorite sights each spring. Alas, some thistles have also made it through the six or eight inches of packed straw. I'm pulling them out as they pop through and will pile on more straw as the potatoes emerge and I can work around them.

After tea I planted out the hardened-off tomatoes, all eighteen plants. The only ones whose identity I was sure of were the Beefmasters I'd repotted; the others got mixed up and lost their labels while I was moving the four redundant Beefmaster seedlings and the two Viva Italias into empty Gurney Girl pots. I also could identify one Gurney Girl and one Viva Italia. So, for the record: three Beefmasters are planted at the path ends of the two tomato beds; the Gurney Girl is at the opposite end of the row in Bed Five, and the Viva Italia is the fifth plant from the path in Bed One. The rest will reveal their identities as they grow and start fruiting.

Viva Italia is determinate; it'll stop growing after a while. The others won't, and will need more room. But since I couldn't tell what was what I had to space everything the same, two feet apart.

This year I'm conducting an experiment suggested by an article in the *New York Times* of December 8, 1991, which Ted saved for me. According to this article, the folks at the Department of Agriculture in Beltsville, Maryland, planted their tomato fields in hairy vetch, in September 1990. The vetch developed a good root system and grew a foot before cold weather stopped it; then in March and April, 1991, it took off again. By April 29 it was four feet tall. That day they mowed it to one inch, and on May 1 the tomato seedlings went in "right into the middle of the hairy vetch. No tilling, no turning, no disturbing the soil. . . . 'we drill a three-inch hole and stick the seedling in.'" The tomatoes planted in vetch in this way did much better than both the unmulched control crops and those mulched with plastic, producing 45 tons of tomatoes per acre compared to 19 and 35 tons. Not for nothing do they call leguminous cover crops "green manure."

By happy chance, when I saw this article I already had most of my garden planted in hairy vetch. I always put in two long beds of tomatoes, about twenty plants. It was a snap to devise an experiment in which half could be grown like they grew them in Beltsville and the other half in the usual way. I've picked main garden Beds One and Five, since Five had the best stand of mulch—not chest-high as in Beltsville but knee-high anyway—and One had been under several feet of leaves all winter and had nothing growing in it at all.

In One I planted each seedling in the ordinary manner: compost and fertilizer in the planting hole, mix into the soil with a trowel, pour in water, pinch off leaves, knock the seedling out of the pot, lay it on its side in the hole so roots will grow out of the stem, wrap the newspaper cutworm collar around it, backfill with soil to the middle of the collar and press down firmly. Over in Five, though, I just pulled circular holes in the heaped-up vetch—they looked like big green bird's nests—drilled a three- or four-inch hole with the bulb planter, dropped in the seedling, wrapped the collar around, and knocked the soil out of the bulb planter to fill in the hole. The soil was wet, and the heavy clay didn't want to come out of the planter. It will definitely be a challenge for the plant roots to push out into that heavy soil not loosened or turned since late last summer, but Beltsville's results are not to be denied. We'll see how it all turns out.

In the late morning the day turned sunny and mid-sixtyish, and I thought it might well be my only chance; so I donned my beekeeping rig (sans smoker) and went out to do what I thought would be a quick operation: put one of the two queen cells from the old hive into the new hive. Only when I opened up the old hive I couldn't find them. I'm sure I remember where they were, but I saw no sign of either cell.

I kept looking for a while, worrying the whole time that the day was too cool for this. Whenever the sun went behind a cloud, I stopped pulling frames till it came out again. Both hives had larvae; but were they the same larvae I put in on Wednesday or eggs that hatched out right afterward? Like the ones I saw then, these appeared to be neither large nor tiny. Had the medium-sized ones of Wednesday been capped by Saturday? As always, it was very hard to see under the cover of bees.

I'll look in again Monday, if I can, but after Monday I'm not going to be lifting anything nearly as heavy as a full super or hive body for a while. If at all possible I'll get at least one new super ready to put on the parent hive before I go Under the Knife.

May 10. In the rain gauge: 2.2 inches in two days! I was going to water, but no need.

I did a few odds and ends today—washed white plastic pots and their black tray, separated double marigolds—but the main project was putting two coats of white exterior latex paint on the five new supers. It looked like rain, so I set this project up in the garage, two long two-by-fours across two sawhorses with the supers stacked on top. Bill Wasch, our contractor friend who built the pergola (among other things), had brought me a can of paint and said to use what I needed and keep the rest till he comes back to work on the bathroom renovation. Everything proceeded without a hitch. The paint's on and dry, and they're ready for the frames whenever I can get *them* ready. Very nice, as Ted said, sitting on one of the remaining hay bales and watching me work, to have a "sort of barn."

May 11. A beautiful spring day, the first in a week. I was out in the garden about 10:30, getting ready to transplant lettuces, when I heard the loud hum of agitated bees. I looked around. At first I didn't see anything; but the sound continued, and I recognized it from the many times I'd opened the hive and knew it couldn't be anything else but bees. So a few minutes later I looked around again; and lo! Back by the cemetery fence, above the jungle, the air was thick with unidentified flying objects. I tossed down my trowel and hurried around to look—and sure enough, about eight feet up in a spruce tree, right next to the jungle path, was a big ball of bees, with a smaller ball right next to it on the same branch. A swarm.

Mine, presumably. Whose else would they be? So my dim-witted "preventive measures" of last Wednesday came too late. *Now* I know, of course, what happened to those queen cells between Wednesday and Saturday: two new queens hatched out of them.

Very anxious lest the swarm take off before I could get ready, I rushed into my rig, got a sheet and the stepladder, pruners and loppers, and lugged all that back into the jungle. I also collected one of the two remaining old hive bodies I got from the Balls, and the last spare bottom board, and took the outer cover off the divide (whose inner cover has a screened-over hole to keep marauders out).

The branch where the swarm had settled was too high for me to cut while standing on the ground, and the jungle growth too thick for the hive body to be set right underneath. It could have been an easier situation for my first solo swarm-hiving endeavor. On the other hand, if I hadn't been outside at exactly the right moment I'd never have known they'd swarmed at all, unless I'd happened to go back there on some other errand and noticed the cloud of flying bees. What snagged my attention must have been the sound of the swarm issuing from the hive, because things quickly got much less noisy. I doubt whether the serious commotion lasted more than fifteen minutes.

Trying to remember how Jim had captured the ancestors of these bees a year ago, I spread the sheet on the grass between the jungle and the former Island, put the bottom board down on the sheet, the hive full of rotten frames on that, and the cover on top. Then I forced my way farther into the jungle and set up the ladder. Realizing only then that I'd need something to catch the swarm in, I went to get a cardboard carton—that's what Jim uses—when my eye lit on a big brown plastic planter, a fake half-barrel, one of two I grew eggplants in two years ago. This one had been covering maple-syrup wood by the fire pit since February. I grabbed that up and dumped out the wood (and rainwater), and up the ladder I went, till I was right beneath the swarm.

After trimming some smaller branches out of the way, I tackled the main problem. The chief difficulty was that I needed five hands: two to cut the branch with the loppers, one to steady the planter, and two more to hold onto the ends of the branch. The result of using only two hands to do a five-handed job was that the first time I pulled down on the branch, some bees fell into the planter sitting on the top step of the ladder and the unbalanced planter fell off, neatly turned a somersault in the air, and dumped the bees onto the ground six feet below. Some flew up, some roiled around where they were. This happened not once but twice before finally, by using the loppers one-handed- and one-armpittedly, I managed to get most of the remaining swarm into the planter while keeping a grip on it with the other hand.

I let the loppers thunk to the ground and backed down the ladder and out of the jungle, clutching to my chest the plastic half-barrel, full of spruce bough and bees. Setting the planter down sideways on the spread sheet, I lifted out the branch and laid it right in front of the hive. And just like magic, in marched the bees.

After watching (and catching my breath and cooling off) for a couple of minutes, I struggled back into the jungle again. The bees that had fallen on the ground were flying up and forming another cluster in the same tree, on a lower, smaller, altogether more convenient branch. Pruning off this one was a snap. I carried the clinging bees out to join the others on the sheet and went back again. There were still a number of bees in the air; others were attempting to cluster on a stick about two inches thick that was lying right on the ground, almost covered by feral lilies of the valley. This I stood upright against a fork in the spruce, to give them a better object to cluster on, but when I came back to check, that little cluster had dissipated. It never re-formed, so evidently the queen or queens were in one or the other of the first two clusters. (Usually one of the new queens kills the rest and takes over the hive, and the swarm goes out with the old queen; but sometimes one or more of the new ones leaves with the swarm.)

Back at the hive, the bees kept on walking in, moving rather quickly out of the spruce needles, over the bottom board and through the entrance. I went inside and called Liz to see if she wanted to take any pictures—I was, wouldn't you know, totally out of film myself—but she couldn't come, so this momentous event is unrecorded by any camera. While still suited up, I pried the inner cover off the parent hive to check the super, found it virtually full, and promised the bees another super before the day was out.

Whew. I got out of my togs, had some lunch, and proceeded with my lettuce planting—four Red Sails, two Black-Seeded Simpson (with two still to go when they get a little bigger and tougher), compost and Vermont Organic Fertilizer in the planting hole, a bowl of water, a cutworm collar, violà. The tiny carrots are just beginning to sprout so I removed the boards and later watered the bed with the little watering can, very gently.

The last thing I did was fit new netting onto the trellis. Ted and I made this structure out of some long pieces of white PVC pipe we found in the garage, along with one elbow joint, when we moved in. We bought another elbow and constructed a trellis of the simplest possible design, by driving two lengths of the pipe deep into the middle growing bed, then connecting them horizontally at the top with an eight-foot length of pipe, using the two elbow joints. Simple as it is, the trellis has proved durable. Winter and summer, there it stands. A year of UV radiation rots the nylon netting it supports, and I replace that every year; but the trellis will stand in the garden till somebody takes it down. The piece of lath I've always used to stretch the lower edge of the netting down has finally rotted and broken, so today I wove a piece of rope through the bottom instead and tied it tight. Thus netted, the trellis looks very nice, ready for melons and cukes to grow on when they come up in two or three weeks.

After checking to be sure slugs still hadn't eaten the growing points out of any cabbages I sprinkled them with diatomaceous earth, a powdery gray substance made of fossilized algae, whose sharp edges pierce the soft membranes of slugs and snails—a great deterrent in dry weather but worthless, unfortunately, when wet.

By the time I'd done all that the swarm had moved in completely and the bees were housekeeping to beat the band. The bottom board outside the entrance was covered with wax, moth frass, and cocoons they had removed and dumped there. My priority was to finish repairing enough frames to fill one of the new supers, so I did that (quickly and less fussily than before), put new foundation in all ten frames, and nailed on the wedges. While I was thus engaged the hospital kept calling with different information—they'd scheduled me for surgery both in the main O.R. and in the Tuttleman Center, I was to come at 10:30, 9:00, 7:00 . . . but finally all the frames had been fitted with foundation and the pins put in. I filled one of my brand-new supers with these frames and took the super out to the parent hive, before going to collect the hive body containing the swarm and move it to its new location.

I'd thought out what I needed to do. I knew I couldn't lift it—bottom board, cover, hive, bees and all. But I could take the hive apart and reassemble it in the cart. I would then—here's the inspired part—trundle the swarm over and *combine it with the divide I'd made last Wednesday*, using the old newspaper trick. A swarm is mostly field bees. The new hive still had no field force beyond a mere handful of nurse bees graduated to foragers, but the swarm combined with the new hive's household force should make a good strong two-story hive, one that would *certainly* contain at least one queen.

When I lifted the hive body off the bottom board, I was amazed to find it nearly an inch thick in frass, cocoons, and nameless detritus. In one afternoon! I had to move the bottom board detritus and all, there were so many bees wing-deep in the stuff. Dumped out in front of the new hive stand it made quite a pile of debris.

Maybe I'm so emotional about the bees, get so upset when I can't handle them skillfully and so cheered up when things go well, like now, because they play such an important symbolic role in my life. It's possible. Early in chemo, when I was utterly downcast and dejected, my good friend Danny Weissbort sent me a book called *Getting Well Again*, by a couple named Simonton. That book turned everything around. The Simontons recommended fighting off recurrence by visualizing cancer cells as bad guys being zapped by ray guns, or as hunks of hamburger eaten by large dogs—whatever had personal meaning for the patient; the ray guns and dogs represented the immune system, first line of defense for anybody with a major illness.

I came up with a beehive. I visualized my body as a vast honeycomb, my white cells as honeybees, and cancer cells as wax-moth larvae and cocoons. I hadn't yet started beekeeping, and some of this imagery got refined as I understood more about the way a colony of bees actually defends itself; but the imagery still works, and I still use it. And seeing the mess this swarm managed to clear out of its hive body in a single afternoon—how could that help but lift my spirits?

When two queen bees encounter one another, they fight to the death; and the worker bees, each loyal to the pheromones of her own queen, will attack any bee in their hive if she smells wrong. Even though my plan involved uniting the swarm with the sisters from whom they'd been separated for a mere week, and restoring their former queen to the nurse bees in the divide, there were still risks if the divide had done as I wanted and started their own new queen.

To prevent fighting between the two populations in case they failed to recognize each other, I took the hive body containing the divide off its bottom board and put on the box containing the swarm. Then I laid a single sheet of newspaper across that box and piled the divide and both supers on top of the newspaper. That way the swarm, which has the capacity to forage, is on the bottom where the entrance is, and the super of honey is up in the divide with the house bees, which are without a means of feeding themselves. In a few days bees from both sides of the newspaper will chew through it, but by the time they do they'll be used to one another's smells and much less likely to fight. If each box has a queen, both sides will accept whichever queen succeeds in killing her rival. My worries ought to be over.

In any case, I've done all I can about this. For weal or woe, the bees are on their own now.

I brought two batches of stragglers back to the new hive, one in the sheet, another that had crawled into the brown planter. All the rest will have to take their chances.

While I still had the ladder out I retied the bee barrier, which was getting saggy. We had our fourth asparagus dinner tonight—the biggest yet, about twenty spears.

(On May 12 I underwent breast reconstructive surgery on my left side and breast reduction on my right. After a night in the hospital, I came home on May 13.)

May 14. I dragged a kitchen chair outside and sat in the grass scraping frames for a while this afternoon, bothered by the occasional bee attracted by the waxy smell but getting four done. Not bad for my second day home from the hospital.

More potato rosettes have poked their way through the straw. The Green Ice lettuce is *slow*—I thought it would have taken off by now; but no, the new leaves are crisp and curly, but they just aren't very big. They need more water than they're getting, obviously, and the dug-in vetch may be causing problems, too. But the tomatoes are already growing well, especially the Beefmasters, those in the vetch looking every bit as big and healthy as the controls.

May 15. In a frenzy of purpose I disassembled and scraped four more frames, bringing the total on hand to ten—eight still in pieces and two already put together. Then I broke out the frame-nailing device again, banged the whole batch together, and put in new foundation. I filled another brand-new super with my beautifully repaired frames fitted with beautiful, fragrant new foundation; then I tucked my jeans in my socks, put on veil and gloves, went out, and gave the super to the new hive.

I didn't want to do any manipulations to speak of—couldn't lift anything, and had bothered the bees enough anyway on Monday—but while the hive was open I did pry one frame out and peer down to see if the newspaper was gone. No sign of it, and bees crawling up and down merrily between the two boxes—so that's one thing that worked out like it's supposed to. Apparently I now have two thriving hives from the same original colony, both foraging cheerfully—the very result I tried to achieve by making the divide!

Busy with all this, I got stung again. A bee got caught in a fold of my jeans and stung me through the fabric. Stinger and all must have pulled out when I jumped because the sting didn't hurt or swell much; it's the least bothersome of the three I've had this season.

Way back in January I ordered brochures from two Pennsylvania poultry hatcheries whose ads I'd seen in *Country Journal*. The brochures came right away, but I stuck them on a shelf after a cursory glance, to await the day when I had time to read up on duck breeds in the two books I'd already acquired: *Ducks and Geese in Your Backyard* and *Raising the Home Duck Flock*. The tea hour today was enlivened by these two helpful volumes. As you might expect, domestic ducks have been bred to specialize in egg *or* meat production, like any domesticated an-imal that does more than one thing well (milk OR beef; mutton OR fleece). Homesteaders, though, need a nice tough duck that yields *both* eggs and meat. There are a dozen breeds of ducks to choose from, but *Duck Flock* describes one variety that sounds just right for us: a multi-purpose breed called Cayuga, after the upstate New York area where the breed was developed. Cayugas are black, flightless, and said to be "One of the quietest and hardiest of breeds." *Quietest* is just what the doctor ordered for suburbia. Cayugas also forage pretty well, and for-aging is essential to my slug-control plan. Mallards forage so well that they can rustle up virtually all their own grub in suitable habitat; but alas, they're said to be very noisy. Anyway, one of the hatcheries offers Cayugas, so that's probably the way we'll go.

One unhappy surprise: the books also make it plain that wa-terfowl put out droppings in tremendous quantity. The smallest num-ber of ducklings you can order from Reich Poultry Farms is six. Six ducks would quickly turn a pond the size of the one we plan into a reek-ing cesspool. The books say ducks can be raised without a place to swim, but that they are consummate water creatures. It's much better for everybody if they can do what they were designed so well by nature *to* do. If we can't put them on the pond, we'll have to think of something else.

May 16. In the rain gauge: 0.3 inches. Carrots still very sparse, potatoes increasing exponentially, one sweet pea (!) finally sprouted in the gar-den, none in the pots. I brought all the trays of seedlings that were still downstairs under lights—eggplants, peppers, marigolds, four tiny Gur-ney Girls—up to set outside; no reason they should continue to be pam-pered down there, big as most of them are, and given that yesterday was the official last frost day.

My sting is red and itches, but there's very little swelling and no soreness. If I couldn't see it I'd think it was a mosquito bite.

I enjoy helping the potato sprouts, with their little rosette-faces of dark green leaves, through the straw—a bit of tender midwifery. They're growing much more vigorously, for no reason clear to me, in the site of former Boxed Bed #1 than in that of #2; and *none* of those "planted" on leaves has as yet sprouted through.

May 17. Rain 0.05 inches overnight. I'm hardening off all the remaining seedlings. Our fifth, somewhat skimpy (eight spears) but delicious, asparagus dinner this evening.

Forgot to mention that about a week before the swarm, when I was standing watching the hive with my back to the bee barrier, a bee flew toward me and over the barrier, apparently carrying a smaller bee beneath it. My first thought was that this was a worker cleaning a dead fellow out of the hive. But I've seen a live bee grip a dead one with great difficulty and barely manage to fly with it from the bottom board to the ground; could a worker possibly fly fifteen feet up with a deader dangling below it? The alternative explanation that suggests itself—though it's hard to believe—is that what I saw was the queen with a drone attached, taking off on a nuptial flight. One book at least says the queen shows no interest in the drones at altitudes lower than twenty feet, and these two lifted off from the hive entrance together; but, as Jim says, maybe the bees didn't read the same books I did.

May 19. A busy, productive day. Pergola report: the hardy kiwis have lots of fruiting laterals and are covered with buds. I tied all the vines that had grown so much they were starting to wave around; I also pressed into service a piece of green wire fencing salvaged from The Island as a support for the clematis vine. This vine is *covered* with huge buds to the top of its post, but can't expand horizontally without something to hold it up.

The hope that a thick organic mulch like straw would inhibit thistle growth appears to be unfounded. I've been weeding thistles all along; when something green begins to push up the straw, it's more often thistle spikes than a potato rosette. But today I pulled up dozens of the little devils. *The straw mulch has not worked.* If I do thistle patrol every single day they won't get out of hand, of course, but you could hardly say I'd *solved* anything. Bummer.

This afternoon I transplanted the four spindly seedlings left in the Burpee greenpack—one Jupiter pepper and three preposterously teensy Gurney Girls—into white pots filled with my homemade potting soil. I then planted out four of the six eggplants, the other Jupiter pepper (much the larger of the two), and one small Gurney Girl seedling which is very little to be setting out, but which will certainly grow faster in the ground than in a pot.

Then I loosened the soil under the trellis, added some Vermont Organic Fertilizer, and planted Amber melons and Jazzer cucumbers. Somewhere I read, and duly noted, that when every other type of melon and cucumber in the author's garden had croaked from bacterial wilt, those two varieties went ahead and produced crops; so this is one more experiment I'm conducting. I would love to identify a variety of cuke that doesn't get bacterial wilt. County Fair '87 escaped wilt in still another writer's garden, but not in mine, though it did produce a sort of crop before dying; but striped cucumber beetles, which are supposed to be (relatively) uninterested in all those bitter-free types like County Fair, were all over *my* vines right along.

I usually wait till around the first of June to plant vine crops (and with our late spring, courtesy of Mount Pinatubo, you might think there'd be particular reason to wait) but the soil temperature two inches down was 80 yesterday, so I thought, hey, go for it. I used all the Amber seed, but there's a Jazzer reserve should one be needed.

The show of carrots is so pitiful I reseeded. Spring carrots almost never work out for us, who knows why. One cabbage has a severely chewed growing point and may be lost; I sprinkled diatomaceous earth on all of them and rotenone dust from a shaker can (years old) on the asparagus, which is being visited by avid asparagus beetles, and comes into the house dotted with tiny black specks: eggs.

In preparation for bean planting tomorrow I dumped four

bucketfuls of wood compost on the designated bed and dug it in. No point in wasting garbage compost on plants that make their own nitrogen.

May 20. ANOTHER SWARM!!!

Ted and I worked together this morning, caging tomato plants in chicken-wire cylinders of divers vintages which have lived outside all winter every year in the dormant garden. Some of these are rusted out and are plainly seeing their last year of active service. To support the cages we use store-bought wooden stakes. Originally six feet tall, most of these have rotted out at the soil line, often not once but several times, getting shorter and shorter year by year until they're too short to support the three-foot tomato cages and have to be retired to blueberry duty and then to peg duty. I dig some of the previous season's rotted-off bottoms out of the beds every spring.

When we finished that job (a couple of cages short owing to retirees), Ted used a hatchet to hack points onto some young trees from our woodlot to make bean poles, something I can't understand why we never thought to try before. They should work *much* better than the store-bought six-foot poles, being both thicker and furnished with branch stubs for the vines to hang onto. Usually, along about the first of August, the entire pole of vines slumps toward the ground like a baggy stocking, dragged down by its own weight and unhampered by the smoothness of the stakes.

Then it was lunchtime. We were sitting in the shade of the white oak with our plates of sandwiches, admiring the stunningly beautiful day, when I suddenly noticed and remarked that the bees seemed very active indeed between the old hive and the new one. As we watched, the air began to fill with more and more bees. I got up unbelievingly and walked over to investigate—and saw that, sure enough, like a recurring nightmare, masses of bees were pouring out of the little opening in the entrance reducer on the parent hive. No doubt about it, that hive was swarming for the second time in eight days.

The good news is, for the second time in eight days, I was home *and* outdoors at the moment of crisis, and thus able to spring into action. I wonder, though, how many times that hive may have swarmed this season and I none the wiser?

This time Ted was home too, another piece of luck. I rushed in for my camera and took a couple of shots of the hive with bees streaming out the front. Then we sat for a while and watched to see what the little devils would do. In minutes the entire garden had filled to the treetops with swooping, milling bees, though none came over to where we were sitting in the shade. They looked like flakes of mica glittering against the huge dark pines.

After a while it seemed they were milling lower and getting denser, and it gradually became obvious they were focusing on the brush at the west side of the yard, next to the garden fence. I got suited up fast and went to check out where they were settling. The cluster was forming in the wall of shrubbery, in a forsythia, only a couple of feet from the ground. The bees made two main clumps—one fairly big and another smaller one, plus various auxiliary clusters here and there in the bush.

I had only one hand-me-down hive body left, with only a few frames in it, most of those a riddled mess of loose wires and wax-moth leavings. But that's what there was, so I piled sheet and equipment in the cart and did again everything I did a week ago. This time it was possible to set up the hive right next to the swarm. I trimmed off some branches that were in my way, then snipped off the branches the bees were clinging to, one by one, and laid them carefully on the sheet. And again they all marched in.

Then I had to think. I don't really want a third hive. These bees of mine are swarming fools, descended from swarming stock; the hive we picked them up from had swarmed already that spring, so my original hive was an afterswarm too, though a good big one. A tendency to excessive swarming is a genetic trait. Requeening is the only way to eliminate it, by breeding it out of a colony. Ergo, I need a new queen (two new queens if I'm going to keep two hives).

But for now I built the afterswarm a hive stand behind the divide, using cinder-block pieces and broken bricks. Then late this afternoon I loaded the afterswarm into the cart, assembled the new little hive on the new stand, and left them to it. Lacking extra covers of either kind, I borrowed an outer cover from the parent hive, substituting a piece of plywood and a brick.

I realized when I picked up the box that this is a much smaller, lighter swarm than the first one. Makes sense that an afterswarm would be, but they'll never make it on their own. As before, the solution will be to combine them with another hive—this time, the badly weakened parent hive they just swarmed out of. If a swarm is put right back on top of the parent hive, the bees just swarm out again; but in a week or so, when the swarming impulse has been thoroughly satisfied, I can pull the newspaper trick and put these bees back where they came from. The parent hive will then have two boxes again. One of my books says you can do this and not to worry about finding either queen—they'll fight, and the best queen will win.

May 21. This morning the new hive was sending forth foragers like the other two. We stood and watched them, Ted holding his mug of coffee. "You have three functioning hives," he observed. "Yeah," I said glumly.

I made up my mind to tear down the parent hive and find out if it had any other queen cells likely to spawn any more afterswarms— one of the books I consulted yesterday says afterswarms leave as virgi[n] queens hatch out and take off on their mating flights. It said a hive c[an] literally swarm itself to death. So right after lunch I took that wh[ole] hive apart, right down to the bottom board.

That's always been a cranky hive, but I have *never* known [it to] be so aggressive. Almost as soon as I'd taken off the covers I regr[etted] not having brought the smoker, and if I'd known how things wo[uld] I'd have gone back then and there and fired it up. Every time I [would] clamp the frame gripper onto a frame, bees instantly covered it [and] hand would be gloved in bees. Every tiniest bump made them [fill] the air with a roar. But I persisted—examined the top box[.]

frame by frame for queen cells, found four, and cut them all. The bottom box was virtually empty, and I removed it.

Then I opened the hive body containing the second swarm. Things were even worse in there than I'd feared—frames without foundation, nothing for the bees even to cling to, let alone draw comb on. They were hanging onto one another and had started to build a little lobe of comb from the top bar of an empty frame. I removed four of these empties and put in four from the bottom box of the parent hive: two frames with honey and two with a little spotty brood. I tried to shake most of the bees off first, but some went in (and were no doubt killed at once). Then I closed up that little hive too and walked back to the house, carrying three of the four cut queen cells and pursued all the way to the door by several furious guard bees.

There was a queen in one cell, pure white and very dead. The other cells were empty, including one exquisitely made cell built on the face of a comb of capped honey (and what it was doing there I can't imagine—definitely not by the book).

All the rest of the afternoon it was as much as my life was worth to go outside. One particular guard bee found me no matter where I was, and dived and hazed and bugged me despite my perfectly unthreatening activities (standing, having tea, planting beans). I had to do the planting wearing a veil, not only on account of this one guard bee but because the forlorn remnants of the swarm were still cruising around at the spot where they had clustered, right by the bean bed.

Other crops and news: the cabbages have been wilting terribly in the hot sun, looking like limp green rags in the bed, but perk up immediately when the shade of the big white pine flows over them each afternoon. We set eight bean poles this morning amid the thin cloud of cruising lost bees, and I planted Romanos around five of them and Kentucky Wonder Wax Pole beans around three.

Yesterday I used a black plastic trash bag, the two flat bottomless boxes of the cold frame, a window screen, and a couple of glass louver panes from Ted's study renovation to make a solar wax melter, a device for separating usable wax from whatever it's mixed with.

Despite the melter's makeshift nature it works well enough. I heap handfuls of scrapings from my frame-repair project onto the screen, and in a while the screen is covered with flattened black guck (wires, frass, propolis) and the pan beneath has yellow wax in it—not much, but I'm still working my way through the bag of scrapings, what bee people call *slumgum*.

Thistle emergence has slowed down (but so, alas, has potato emergence). One Green Ice lettuce died of unknown causes. Asparagus dinner #6 tonight.

May 22. This morning I dug compost into two more back-garden beds, and this afternoon did the same to the triangle bed. We're down to about a third of a big garbage can of overwintered compost and maybe three-fourths of a smaller can of wood compost, but the whole garden's dug except for the future sweet-potato bed. I've measured out this treasure pretty well.

IMPROVISED SOLAR MELTER (CUTAWAY VIEW)

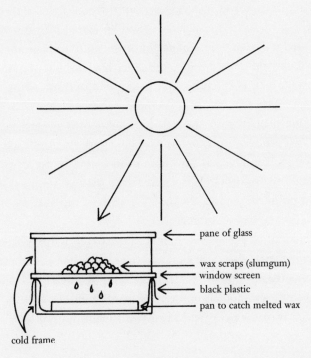

pane of glass

wax scraps (slumgum)
window screen
black plastic
pan to catch melted wax

cold frame

It's suddenly summer and boiling hot in the middle of the day. I never used to wear a hat while gardening, but now, mindful of the thinner ozone layer, I always do. And I try to work outside before 10:30 and after 3:30 when the sun's this bright. The solar melter has melted almost my entire hoard of slumgum.

With dinner we ate some lettuce I picked this morning in a salad with some marinated two-year-old snow peas that had surfaced in the freezer. A mistake: the snow peas were awful and plentiful, the lettuce sparse and understated. An instance of how my reluctance to waste one precious organic scrap of garden produce can be carried too far.

I hauled the big blue watering can out and watered lettuces, eggplants, and the trellis furrow, then refilled and gave these crops a second drink. Not good enough. The trickle lines need to be set up. I'll get to that in a day or so.

There are still two trays of seedlings on the patio, getting bigger, and these have been discovered not by just the odd thirsty bee but whole squadrons of bees; bees were zipping back and forth all afternoon, drinking water as it percolated up through the soil in the pots. They must crave water in the heat—God knows where they get it otherwise—and tell each other about this pitiful source. Tonight, after bee bedtime, I carried the trays out to the garden in the dark. We have to have a usable patio. Ready or not, these seedlings are going into the ground.

One success from the period just before my surgery on May 12 was that I'd learned to work steadily for long periods of time. I'd stopped rushing at a job and wearing myself out in the first hour or two; I was getting up in the morning, having breakfast, then getting into my work clothes, shouldering my spade and rake, and stumping off to the fields for long, steady, productive days.

Now, after the surgery hiatus, I have to learn this lesson all over again. I can't settle down properly to the task at hand. The steadiness is lost. I'm sure it'll come back, but the gaining and then losing focus because of interruptions is one way of understanding why this year as I conceived it originally—and ideally—is beginning to look impossi-

ble. The trouble is mostly that unless I keep my nonagricultural en-
deavors—writing, teaching, relating—in the air throughout this year,
I'll have an immense amount of catch-up to do when I return to ordinary life. But also I can't ignore them now because I have to meet my
half of expenses, and homesteading—for all its tangible and intangible
rewards—hasn't produced a cent of income so far.

The need to stay solvent is one factor that interrupts my concentration; the surgery—and postsurgical loss of focus—is another.
Add to that the effect of the many conversations we've had of late about
where to live when Ted retires. Rose Valley is out, we can't afford to
stay here; but even if that weren't true I feel more and more clearly
that, for all its beauty, this *place* isn't what the Homestead Year is about.
I'm like a sharecropper or tenant farmer, appreciating the splendors of
where I am but certain in every cell that I do but sojourn here, that this
is my work and my design, but it isn't my land, isn't my home. I find
myself beginning to think we can do without the greenhouse and ought
to trim our ambitions about the pond to better fit a temporary future
here.

Trying to have my homestead year has shown me that I can't—
can't, anyway, have it the way I wanted to. All I've really been able to
do is shift the emphasis. No wonder I feel glum.

May 23. I'm just absurdly emotional about the bees! Let them act like
they don't like me and I'm genuinely upset; let me look into a hive and
get good news, and my whole day improves. This morning I opened the
afterswarm hive, expecting to find things in the tatterdemalion state in
which I'd left them last Tuesday, but behold! The bees had cleaned up
and repaired things to the point where I couldn't see any point in replacing their frames with new ones. What a lot of hard work they've
done since Tuesday—only four days! It makes me hopeful that things
will go well *next* Tuesday, when I'll add this afterswarm to the parent
hive using *two* sheets of newspaper, and may the best queen win.

Another sunny day, with temperatures up in the nineties,
finished up the last batch of comb in my solar melter. I removed the

screen from between glass and pan for an hour and let the sun stream straight through and melt the wax completely, then strained it through several thicknesses of cheesecloth. Lots of guck strained out, but lots more got through; there's bound to be a better way. We dunked some short pieces of string in the pan to make wicks, then poured the liquid wax into two custard dishes and set the wicks in the middle, held up by little sticks till the wax could harden. When it first began to solidify on the tongs and glass dishes, it looked beeswax yellow, but as it cooled it darkened, and the finished product is more tan than yellow. Some of the wax I scraped off the frames was very old and dark; it's newly secreted wax like honey cappings, with no impurities, that makes those beautiful yellow candles you see in boutiques.

After dinner we lit one of ours to see what would happen. It flamed up beautifully, but when the wick burned down to wax level it went out—which illustrates why a taper is the shape it is (so the wax will melt and run down out of the way of the wick as it burns down). The whole house smelled of beeswax for a couple of hours afterwards. I must get some candle molds and some instructions!

The bees came looking for the trays of seedlings and hung around grumpily trying to make do with the potted hydrangea on the patio (which hasn't been watered in a coon's age) before giving up.

Cabbages, tomatoes, and potatoes all went limp as usual during the blistering day, so I waited till after tea, when the garden was in shade and there was no risk of tearing limp cabbage leaves, to lay out the trickle lines in the main garden. I don't like the system I bought the first year we were here, but it's set up to fit these beds and it still works and I've already got it, so there's no thought yet of replacement. But I'd like to get soaker hoses with perforations over the whole surface, not these annoying little emitters that so often aren't just where you need them to be. After no little cursing and finicking, I've got the main garden system all in place. Not a moment too soon, either; the soil has become very dry. Great for digging yesterday; not so great for plant growth. They've been slurping it up out there since 6:00 and it's now

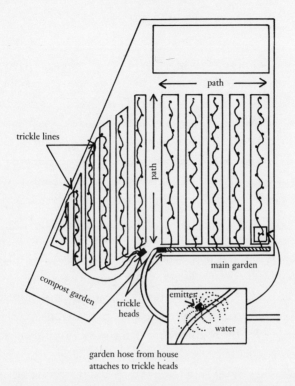

trickle lines

path

path

main garden

compost garden

trickle
heads

emitter

water

garden hose from house
attaches to trickle heads

almost 10:30; I'm sure they feel a lot more comfortable. I don't want to lay out the lines for the back garden till more of it is planted.

After dinner, bees calming down for the night, I went out with the stepladder and tied up the bee barrier again; one side was sagging badly. This time, instead of tying it around the top, I poked the cord through the netting and pulled the knot really tight. Should hold much better, but may tear through. Never mind. In my opinion this is one great solution I devised.

May 24. I planted out the four hot pepper and one (very small) Jupiter green pepper seedlings, tied up some rambling kiwi vines, and picked asparagus for our seventh and final asparagus dinner; but the big project was the trickle system for the back garden, and now that's all been seen to. When it was set up I watered for a while, but then a front came through followed by rain, so our immediate needs have been met.

Both last night and tonight we went out on slug patrol, with a flashlight and a bucket of water. Three of the cabbages have had their growing points eaten out by slugs, the big orange type of slug we raise so successfully here, though these aren't all that big as yet. But big appetites. We hand-picked—or tong-picked—a couple of dozen, and it did seem to make a difference. Soon these cabbages will be big enough so the slugs are no longer a concern, but not quite yet. We must keep patrolling. This is the sort of job that goes much better with two: one to turn over leaves and shine the light, one to carry the bucket and wield the tongs.

On the kiwis, lots and lots of flower buds like little stalked knobs. How pleasant if they should bear this year! A few tiny currants on the three bushes, too, which are growing very nicely.

Ted has been beavering away at the jungle, managing it like the tiny woodlot it is, and has stacked up an impressive pile of small rounds for next winter. He throws himself into the work, is unselfconscious and totally absorbed by it.

I've enjoyed these hot bright days and the routine they suggest or impose: outdoor work in the morning, lunch, deskwork, teatime, second outdoor stint as things cool down, shower, dinner, evening off (record-keeping apart). Routine that integrates farming and essential nonfarming activities is just the ticket. But now we're behind a front and temperatures have dropped 40 degrees in a single day, so for the present it's back to chill and damp.

May 26. Wet weather—no chance to combine the hives. While Liz was here for lunch we took an umbrella and went out to tour the garden anyway. One of the hot peppers I transplanted yesterday had been broken against its cutworm collar by the high winds—not hardened off enough for so much bluster, alas. At least it wasn't one of the green peppers, in higher demand and shorter supply.

Tonight for dinner our first real lettuce salad.

May 29. Spoke too soon: today I found the littlest green pepper listing against the top of *its* cutworm collar, broken, too.

It wasn't really warm enough but day before yesterday I decided to go ahead and put the parent and afterswarm hives together. Simple enough to do. Neither hive was open more than a couple of minutes and should have had no problem warming up again quickly afterward. I waited till early evening, when most of the foragers were home—so that as many bees as possible would get moved—and used two sheets of newspaper instead of one.

After today's inspection and maneuverings, the beeyard looks like this: one two-story hive with one super (the afterswarm/parent hive combination); one two-story hive with *three* supers (the divide/first swarm combo); and one set of cinder blocks with nothing (please God) on them. The newspaper trick has worked again, and all seems well.

No idea what's going on with queens and brood and all that, and no way to check without lifting too many supers off and on. Or rather, I can check the parent hive easily enough—and probably should, in a week or so. But not the swarm hives; those many pounds of medicated honey have now effectively sealed me out of the hive for the rest of the season. The two queens in the parent hive will have fought by now, and one or the other will have triumphed and taken over. They're this year's queens, both of them, with no differences I'd be able to detect—even if I could find either one. Finding the old queen and killing her matters only if you've just paid good money for a new queen from a breeder and want to ensure that she—and not the scruffy one (who might win a battle if it came to that)—takes over the hive. If you don't care who wins, you can safely leave it up to the bees.

Beework wasn't all I did today. At long last I cut the black plastic sheeting to fit and covered the back garden with it, stapling it down at crucial places with a packet of earth staples I must have had around here unopened for years. When the plastic was installed I spread straw over the path parts, cut X-shaped holes with a knife in the parts covering the triangle bed and last compost bed, and planted the Butterfruit corn in the holes: hills on 12–14 inch centers, seven seeds to a hill, with intent to thin to three or four. I threw a handful of Vermont Organic Fertilizer in each hill. Thank God that's done! I hate putting all this plastic in my garden, but nothing else—including, as we've learned, two feet of straw mulch—will keep Canada thistle down. Except newspapers, which, while tolerable in the paths, are a mess in the beds, and don't last the season anyway.

Several burstingly healthy Rote Erstling potato plants had been wilting for the past few days. I thought at first it was the heat and sun, but they stayed wilted in cooler wetter weather too so I finally dug down under the straw and found a *mouse tunnel* winding right underneath the roots, between the plants and the soil. Below the roots was only air.

I pressed the roots down into the soil and tried to replant a few plants, wondering what you do about mice and thick straw mulch, and why I hadn't read about this problem anywhere. Then I remembered the bottle of outdoor animal repellent that came from Gardeners' Eden with the gift certificate they sent me. It's supposed to work on squirrels, so I figured it ought to repel mice too, and sprayed it liberally around the area of former boxed Bed One. Nice lemony smell. I hit some leaves, but it says 100% natural ingredients on the bottle so I don't suppose I've injured my organic status here. Maybe it'll work. We'll see if more potatoes wilt.

Today I found and brought inside the first bird-pecked, half-ripe strawberry.

The dogwood trees between us and the Purbricks have been slowly dying for years, from the bottom up, of that incurable dogwood disease. Last year Ted cut down one of them and turned it into firewood; today, since the bloom has ended, he felled another, cutting a notch with the electric saw on one side and making a straight cut from the other. He dropped the tree across the driveway, bucked it up using his new sawhorses, and had sawed the whole thing into firewood by afternoon. Dogwood makes wonderful firewood.

May 30. We noticed that one of our white pines and one of Jim's have a single browning limb, and I wondered aloud if the currants, despite Mr. Miller's assurances, could possibly have infected the trees with white-pine blister rust already, six weeks after being planted. It seems impossible, but in the middle of last night I woke up very worried about it, and got up and reread the *O.G. Encyclopedia* entry. It plainly says: "Consort (rust resistant)" and "plant resistant varieties or site 200' from the nearest white pine." Jim's tree is that far away, though ours is not. I worried about this for quite a while, probably needlessly. I'd call somebody about it, but unfortunately no "authority" really seems reliable.

As we were driving along this afternoon I kept a sharp eye out for other white pines that looked droopy. Some do, some don't. Pines do die from lots of causes besides blister rust. "Fatal cankers," says *OG*. Cankers are diseased areas of woody tissue. Maybe somebody can at least describe for me the course of the disease.

Rain all day. In the gauge: 0.03 inches.

May 31. Again, rain all day: 1.85 inches. On slug patrol we plucked a dozen or so big slugs and some babies off the cabbages, which have grown phenomenally in this heavy rain. One whose growing point had virtually disappeared has bounced back with a frill of little leaves. The lettuce also grew like crazy; we should get a big salad tomorrow as well as the really final serving of asparagus for dinner—I won't cut the spears any more this year; it's time they got on with their growth. cycle The first beans have sprouted. A couple of Beefmasters are sporting big fat flower buds. Thistle patrol tomorrow needed desperately; thistles love rain!

June

June 1. A full gardening day; also the day of my fourth sting of the season, and only myself to blame. I was hoeing the weed-choked asparagus bed, chopping down the row vigorously and forgetting how close I was to the beehives. Suddenly a bee got caught in my hair. She escaped but zipped back and stung me on the jaw anyway. Remembering what my friend Vicki had told me abut Epsom salts, I went inside to flick the sting out, and find that box of Epsom salts I bought years ago after reading that some crop or other thrived on the stuff. But I could not remember exactly what Vicki had said to do with it, so I just mixed some with enough water to make a paste and plastered the paste on my jawbone over the sting. This evening, no swelling and almost no pain—and this was a full-voltage sting.

Years ago, when the apple trees were babies, I practiced a form of protective overkill on them all: white exterior latex paint on the south side of their trunks to protect them from winter sunscald, a ring of hardware cloth to keep mice from nibbling the bark in winter, a two-foot circle of fencing to keep rabbits out, and a chicken-wire ring on the fencing for the same reason we applied chicken wire after the fact all around the garden fence. Five years later, grass and suckers and various other things had grown up thick enough to choke the space enclosed by the rings of fencing. So this morning I removed these rings from all five trees and cleared out the mass of vegetation inside, in preparation for mulching them with straw out to the drip line.

Everybody says apples are very hard to produce organically, at least to commercial standards. Checking out the little apples, I saw with sorrow, but without surprise, that many show signs of insect damage. So it goes. I picked off and squashed a couple of gypsy-moth caterpillars, but this is management at the level of trees. At the more difficult level of flowers and fruit, I blew it. My plan to monitor apple pests never got beyond spraying dormant oil.

Forsythia has invaded the blueberry hedge, like always. I cut it out, right to the ground. When we set up the bird-netting system in a couple of weeks the hedge will be ready.

This afternoon I dug the rest of the wood compost—about five buckets' worth—into main Bed Four, next to the Beltsville experiment, and planted the sweet potatoes. The soil was wet and heavy, but I wanted to get the plants in the ground. Sweet potatoes are supposed to be planted in a ridge ten inches high and a couple of feet wide, exactly the shape of the raised beds. I did as the instructions said: made a furrow down the middle of the ridge, separated the plants, which were tangled up together, stuck in the trowel ("make a drill"), poked in the roots, and pressed the drill shut. Do this twenty times, and your sweet potatoes are planted.

My order from Jung was for a "sampler" of four plants each of four varieties that do well in the north, but in fact they sent a few extras and I ended up with five Vardaman plants, four Jewel, five Georgia Jet, and six Centennial, for a total of twenty. No fertilizer; just vetch and compost. Judging from the tomatoes, though, vetch is as good as fertilizer and maybe better. So far, whatever minor observable differences there are are in favor of the Beltsville bed.

I picked more lettuce and some half-ripe strawberries I was afraid the birds would get if they were left to ripen fully on the vine. Doesn't matter anyway—strawberries color up beautifully right in the kitchen. I'd planned to mulch with straw to keep the berries from lying directly on the ground and rotting in wet weather, as has happened in other years, but there are still a lot of oak leaves in the bed and I decided to skip it—plenty to do without make-work.

June 3. Finally, a beautiful warm sunny day. I grimly worked down my list: replant cukes (the first-planting seed rotted); plant out marigold seedlings where space permits, plant the poor suffering Gurney Girl seedlings with their malnourished purple leaves and let them sink or swim (two went in the vetch, one at the end of the other row); repot the pennyroyal seedlings in individual pots; attempt to save the broken pepper by burying it up to its neck; install a trickle line in the sweet potato bed; plant popcorn; prune back the roses wherever they're waving long stems full of thorns in everybody's way. I crossed off and crossed off.

All afternoon I worked on the back bedroom where gardening supplies get stored till the next time they're needed, and which is to perform its other role of guestroom this weekend. Then I picked strawberries, a cheerful daily task to look forward to for the next couple of weeks. Not many today or yesterday, but enough for Ted and me to have with some Product 19 and milk: delicious! All through the season I keep two bowls of strawberries going: one of yesterday's pick, waiting an extra twenty-four hours to get fully ripe, and one underripe bowl of today's, salvaged before birds or slugs or fungi can do their various sorts of damage to the berries. A good day of steady work and accomplishment, morale building and soothing.

The sweet potatoes look pretty poor; they should have been put in water immediately, not a day-and-a-half after they arrived. Two of the smallest probably won't make it—but I won't pull them up till I'm sure they're well and truly dead, this being one very tough plant by all I've read. The Krantz potatoes and one other type are forming buds, getting ready to flower, but I think every time I go out there how disappointing the showing is this year.

Till now I've been too busy with more urgent projects to give the Edible Fishpond much focused thought; but now that the spring semester's over, and Ted's through teaching, its number has finally come up. The past couple of days I've been looking at water-garden catalogs, comparing prices and offerings. The Paradise Company has the most attractive prices and sells a collection of plants and planters without fish. That suits our needs better than the Lilypons "collections," which are pricier and come with preselected ornamental fish and plants; I aspire to catfish, not koi. Beauty is as beauty does.

It *is* my impression that Lilypons is the leader in the water-garden biz. This impression is derived partly from Liz, whose own pond and its trappings came from there, and is reinforced by the fact that Ted and I once stopped to visit the Lilypons store in Buckeystown, Maryland, on our way home from a weekend in Washington. Closeness, plus the sense we acquired of its being a real place, with ponds and plants, incline me to sway the vote their way; but price dictates otherwise. Right now, though, Lilypons is having a sale on flexible pond liners, so at least we can buy that from them. The other stuff's not cheap, but even at a sale price the liner will be our biggest expense by far.

We laid out the perimeters of the pond with a clothesline as best we could on the bumpy, stub-fraught ground, and once Ted saw how big a 16′ pond would actually be, he decided that my original 16′ X 7′ or so dimensions were about right. Finishing clearing the site is the priority now; even if we don't put ducks back here, it's still the best place for a pond. Plus I need to order the liner, if the pond's to be excavated and filled before we leave on June 13 for our time-share week at Shawnee-on-the-Delaware.

June 4. I went to Frank's Nursery this morning and bought four green pepper seedlings (a type called Better Belle Improved), eight brussels sprouts (hard to find healthy ones, but these look okay), three 3′ X 50′ rolls of miracle mulch for the electric fence, and two bags of potting soil for the two eggplants I'm going to containerize. Set on the patio step next to the new seedlings from Frank's, my eggplants look *awful*: yellow, speckled, feeble, and sickly. Frank's had perfectly good eggplants too; pointless to have grown those from seed when there's nothing at all special about the variety, and when I'm not honestly that concerned about their nonorganic early infancy. I should have just waited and bought both kinds of peppers and eggplants, and marigolds, as well as sprouts. For the future: START FROM SEED ONLY THOSE VARIETIES THAT I REALLY WANT AND THAT CAN'T BE OBTAINED LOCALLY AS SEEDLINGS, tomatoes and cabbages mostly. Plus lettuce, since lettuce can't be direct-seeded here in Slugtown.

I opened both hives just long enough to see what was going on in the supers. Not much doing with the parent-hive-cum-afterswarm, whose single super is still only partly filled. The swarm-hive-cum-divide is much better—virtually all the comb in the top (new) super drawn, much of it filled, and some even starting to be capped. Come what may, we should have *one* super of honey to extract this year. Next time I need to take along the hedge clippers and trim back the mock orange, which is now picturesquely in bloom but which has overgrown the swarm hive to the point where it's hard to work it.

I picked a healthy quart of strawberries this afternoon, just ahead of the rain. Saw a catbird perch innocently on the frame around the strawberry bed before hopping down into the berry patch. I scared it off and went for the bird netting. A timely move; while spreading and tucking it under the edges of the frame, I noticed a fair number of pecked berries.

June 5. A rainy day, just the ticket. With our bucket of soapy water and tongs, Ted and I went on slug patrol among the cabbages, potatoes, and lettuce, and picked off jillions of those humongous orange slugs that love cabbages. They were also doing a number on the potatoes. Too bad the ducks won't be coming for another five-six weeks!

The mulch method just is not working, that's the truth about that. Thistles poke right up through it, mice tunnel between soil and plants and kill the plants, and slugs are provided with the perfect environment: always wet at soil level and lots of hiding places. All that can be said in its favor is that it does conserve moisture wonderfully, so I don't need to water the potatoes ever, and that it keeps down weeds— but not perennial weeds.

The sweet potatoes are looking better. I'm sure I was right about the two deaders, but the rest are starting to stand up in the rain and take notice. One corn possibly up; didn't look closely in all the holes in the plastic mulch (which anyway are full of sprouting nut sedge) but this rain should do the trick. Both types of beans are well sprouted and looking great, as they always do. So do the tomatoes in both beds, by the way—bushy, healthy, terrific; you'd never guess they suffered from Purple Leaf Underside deficiency as tykes.

Ted and I have been made an offer we can't refuse from our time-share exchange network: a week at any resort with available space for $89!—so we'll be off to Powder Ridge, in the Utah Rockies, in July. Our absence at that busy point in the season alters the homestead timetable somewhat. I called Reich Poultry Farms today to find out whether Cayuga ducklings would still be available in mid-July. The woman I spoke to said that was pushing it—how about late June? "We'll be away off and on until July 17," I told her. She said she'd contact the people who raise Cayugas for them and let me know. If the answer's no, I guess we fall back on Mallards and view the extra quacking as a trade-off for the extra foraging zeal.

June 6. My old friend Rika Lesser is here from New York for an overnight visit, but we labored through the afternoon anyway. I planted the four-packs of peppers and brussels sprouts from Frank's, the latter in compost Bed Four, and settled the two ravaged eggplants in plastic planters filled with yummy-looking potting soil. (One of these was what I used to hive the first swarm, and before that to keep maple-syrup firewood dry. Reduce, reuse, recycle.) While I was engaged with all this, Ted was clearing the former Island, site of the future fishpond, of its stubble of forsythia, honeysuckle, assorted roots and stumps, iron pipes, etc. Next week is pond week, come what may. This morning I called Lilypons and ordered the liner. Then I called Paradise. When I explained that we'd be away week after next, I was told to wait to order the plants and snails till later; so that's what I'll do.

June 7. This morning Rika and I mulched the last four paths, she laying the newspapers, I following and spreading the straw. It was hot and humid, the sort of weather she hates, but she held up well, and now the whole garden is planted and mulched: hooray!

I'm sure that being here feels as alien to Rika as visiting her in Brooklyn always feels to me. She hasn't said so, but a certain determined grimness in her aspect speaks volumes. Our long friendship is founded on shared literary enthusiasms and activities, both being poets and translators of Swedish poetry—*the* translators of Swedish poetry we've both been heard to say, without hesitation and with deep mutual regard—and we're close in other ways as well. But while these literary enthusiasms draw me regularly into Rika's world, the urban sphere where she feels most at home, nothing in her life draws her into nature wild or cultivated, where I'm always wanting most to be. Yesterday when I made some remark about wishing I were doing all this in the country, she looked startled and said, "*This* seems like the country to me."

I once made a list of all the writers I could think of who care about, write about, and practice, agriculture (they had to do all three): Wendell Berry, Noel Perrin, Donald Hall, Ted Hughes, E. B. White. A short list innocent of women, unless maybe Maxine Kumin could be counted in on the strength of her gardens and horses. So I—not Rika— am the oddball, the one whose discordant passions require a continual shifting from one mental set to another to another, trying to have it all and never getting quite enough of any of it.

For all that she hates bugs and worried about them from the minute she got here, Rika did watch from a safe distance, through field glasses, when I opened the parent hive this afternoon. And ironically enough she was witness to a great insect event, because O joy! I finally saw the queen. I pulled a frame and there she was, in the act of backing into a cell to lay an egg, and apparently unfazed by the sudden exposure to light. The shiny black bald spot on her thorax would have identified her even without the egg-laying behavior. The brood nest, where she was laying, is mostly in the top box, suggesting though not establishing that this queen, winner of the skirmish to the death, was the queen of the afterswarm. Whichever queen she was, it was a pleasure and a privilege to meet her.

The hive looks to be in good shape generally. Two frames are filled on both sides with capped brood—a beautiful sight—and there were some tiny larvae right at the top of one frame in the bottom box. And the hive is *bursting* with honey—much more than the bees ought to need to get them through next winter.

Through the afternoon Ted went on working to clear The Island. Tonight there's one big stump left to dig out, and some smaller stubs, but it's a huge job accomplished all by himself in only two days of work, which augurs well for when we work together. The pond is the one project I secured Ted's agreement to help me with before the Homestead Year began. He's not unwilling, but *is* anxious about exactly how much sheer labor and time the excavation is going to require, and tends to be short with me about it.

I called Bill Wasch tonight and arranged to borrow a four-foot level; everything I've read implies that this is the one indispensable tool. The liner's supposed to be delivered in the middle of next week. If the weather cooperates at all, we should get done before leaving for Shawnee.

I saw a white cabbage butterfly fluttering attractively over the cabbages at lunch time. Later inspection revealed a number of eggs on one of the biggest cabbages; Bt time may be upon us once again.

June 8. At two minutes to nine this morning, as I glanced outside, the corner of my eye caught a movement: a baby rabbit on the grass path inside the closed-up garden! This is a self-correcting problem; since rabbits grow like weeds, in another week or so a baby this size will be too big to get through the fence. But they can do a lot of damage in a week.

I yelled; Ted went out and chased it till it escaped through the fence back by the compost bins. We discussed setting the Havahart trap, but didn't have any carrots and couldn't think what else a baby rabbit might prefer to fresh lettuce. I had to catch a train, so we let it go. This evening I found the freshly transplanted brussels sprout seedlings seriously chomped on, to the point where I'll probably have to replant a couple of them. But we still have no carrots, so the Havahart still hasn't been set.

While I spent most of the day in Philadelphia, Ted was spending it digging out the one remaining mammoth stump—two stumps, as it turned out—left at the pond site. I came home at teatime to find him exhausted and filthy, but successful. (Not happy, however. He hated the whole job.)

After tea I changed and went out with my pruners to cut back the azalea next to the pond site, which has finished blooming, and clear out the tangle of honeysuckle at its base. In a couple of hours I had lifted the skirts of the former—pruned off low-growing branches, trimmed, cleared out all the deadwood—and removed the latter where I could reach it. This included pulling up a lot of roots. But the main stem of honeysuckle seems to grow right out of the center of the azalea stems, and can't be eliminated without eliminating the azalea, too, so we'll have to settle for a little judicious snipping from time to time. I also created some breathing space between the azalea and the boxwood planting. Now we can work on the pond without stiff little azalea twigs poking into our eyes and ears.

June 9. The pond excavation is under way! We went out first thing this morning with stakes and string and the metal tape measure, and staked out the site: 18 feet by 12. Just as we were starting this, Bill Wasch and his son Jeff turned up with the level and square. For a couple of hours after that I hauled away the dirt Ted was digging up, and dumped it in three piles (topsoil, mixed soil, clay). Eight shovelfuls makes the biggest manageable load. We're using the little plastic cart, which is very light and easy to overturn; the garden cart isn't the cart of choice for this dumping work.

When I had to leave for town, Ted carried on alone with the digging *and* dumping. By the time I got back, he was down to grade in a hole about four feet by seven. Terribly heavy labor for a sixty-nine-year-old man, however fit and youthful! His procedure was to loosen the topsoil with the pick, scrape it up with the shovel, pick again, shovel again, layer by layer down to grade, and course by course from one end to the other . . . he worked on a road-building crew in New Haven during the summers when he was a young professor at Yale, and the skills are similar and very useful, but taking it slow and steady now is essential. If either of us pulls a serious muscle or gets injured some other way—Ted has broken blisters on both hands, for instance, not good—we're in trouble. The soil is wet and very heavy. We must both be careful.

The garden continues on its own course for all our preoccupation with pond building. The popcorn is up. Five Jazzer cucumbers are up also, and big enough that they must have sprouted several days ago. Only five, and big as they are, that seems likely to be all. The sweet potatoes have *all* taken hold, including the two I would have sworn were goners, and are producing new leaves. Both hardy kiwis have suddenly burst into full and gorgeous bloom.

While admiring the pond excavation I heard a suspicious rustling in the garden. Sure enough, the baby rabbit was in there again. Ted chased it out—through a different place this time, by the compost holding cans—while I went to inspect the damage. More brussels sprouts had been chomped; it's obvious that unless we do something, there won't be any more in another day or so. I went inside and rummaged around till I found a big old piece of used Reemay—what the catalogs call a "floating row cover," gauzy and semitransparent—and draped it over the sprout seedlings double-thick, weighing each end with a stone. It will ward off the bunny, who will be growing daily a little nearer toward being too big to fit through the fence. He seems to have nibbled on a few beans and cabbage leaves too, but obviously prefers sprouts to the other offerings. As far as I can tell the lettuce is untouched.

Everywhere I haven't pulled it off or out, honeysuckle is blooming and sending forth my favorite of all flower fragrances. A honeysuckle tangle has covered the forsythia bush outside the bedroom window. All night the scent drifts in. If it bloomed for a longer period— say a month, I might very well let honeysuckle go ahead and strangle almost anything it wanted to grow on. The Swedish verb *dofta*—"to give forth fragrance"—always comes to mind at this season. Such a scent deserves its own active verb.

June 10. Through the kitchen window this morning I saw a honeybee foraging happily in the kiwi blossoms. It would be great if these two-year-old vines would bear a crop this year. In earnest of my husbandry, before girding up my loins and going off for the dirt cart I spent an hour tying loose tendrils to the pergola rafters.

While thus engaged on top of the stepladder, I heard a droning and turned just in time to see a very large bee—larger than a bumblebee—disappear into a neat round hole in the bottom edge of a pergola rafter! Bill Wasch has been telling us about what he calls "barn bees," that make a hole in barn boards exactly ⅗ of an inch in diameter. Obviously this is the same kind of bee. Not too good for the rafter, but how bad? How far in does she tunnel, and how much does that weaken the board? I should squirt some rotenone up there if a barn bee is a kind of glorified carpenter ant, but out of feeling for bees of every sort, I hope it won't be necessary.

A quick check of the garden revealed more wilting potato plants, beneath which were more mouse tunnels. I made a decision I hate to have to make (but if I don't we won't get a crop at all): to pull off the straw and cover the plants with topsoil from the pond excavation. When I removed the straw it proved to be full of fat orange slugs—no wonder the lower potato leaves are lacelike and the marigolds completely defoliated! Ted plans to cut the grass before we leave for Shawnee; we'll mix this partly decomposed straw with grass clippings and have super-duper compost in no time (and steamed slugs).

I started on this potato project, but doing two major jobs at once is just too hard. I settled for dumping numerous cartloads of topsoil on a plastic shower curtain next to the potato bed, and going back to carting only. I hauled and dumped for hours, personally moving and dumping tons of earth (which Ted had personally dug up and shoveled into the cart), lifting, heaving, straining like an Egyptian slave building a pyramid, trying all the while to be careful not to pull a muscle.

It took us a while to figure out a good system for working together. Ted didn't like to break his rhythm, waiting for me to get back with the cart, but he hit upon a scheme of shoveling into the cart when it was there, and into a pile on the ground when it wasn't. That worked fine. Or sometimes he wielded the pick while I hauled and dumped. Ted had started at 8:00, I about 9:30; we kept at it all day, stopping just briefly for lunch and tea and one or two other short breaks, and quit about 7:00, having agreed that there was nothing magical about the recently revised goal of fourteen feet either, and we could just as well stop at thirteen if we liked. We'd planned to keep working till the day was over, and stop wherever we'd got to; and thirteen feet was it.

The hole is neither perfectly level nor perfectly square, but Bill's square and level made it more nearly both than otherwise. Its bottom isn't really flat, either. It has one marginal shelf, also not very flat, and its edges slope at different angles. But so what? The deepest part measures over two feet and the shallowest somewhere around twenty inches (another compromise agreed upon about halfway through: that the bottom could slope upward to make the far end shallower than the near end). None of these imperfections should make the slightest difference to liner installation, plants, or fish.

All the time we were working the two delicious fragrances of honeysuckle and honeybee hive (nectar and new wax)—the latter wafting clear from the other end of the yard—kept washing over us.

I'd been told that Jack's Firehouse, a Center City restaurant, keeps live catfish in tanks to be served to the clientele, so during our lunch break I called to see if they would sell me some catfish fingerlings. Jack's rerouted me to Hopkins Seafood. The catfish man there told me they buy fresh *dead* catfish, which are trucked up here from Mississippi and which he supplies to Jack's. He said there are no local growers.

He gave me phone numbers for the Catfish Farmers of Mississippi and the Catfish Farmers of America and suggested I call to get a price quote for an air freight shipment that I could pick up at the airport. Good suggestion. I'm also going to try the aquaculturist who directed the fish-farming project at the Rodale Research Center, and who still lives and works in Kutztown. The Paradise catalog lists blue and golden channel cats, 3–4 inches, each $4.50 plus shipping. It appears the catfish problem isn't going to be the snap to solve that I had imagined. But I'll solve it—it just may take a little longer and cost a little more than expected.

June 11. This morning I took a clipboard and went around the garden and yard, taking stock.

Tomatoes: The first plants are flowering, two big Beefmasters in the vetch and four other smallish plants, one in the vetch row and three in the control row (I'm guessing Viva Italia, which, being determinate, finishes growing and starts flowering before the indeterminate types that keep on getting bigger all summer long). The big vetch Beefmasters are at the tops of their cages.

Cabbages: No sign of the eggs I spotted on Sunday (June 7), nor of little worms nor damage. Some minor rabbit damage, not bad; but *all* the cabbages, those with badly slug-chewed leaves and those whose outer leaves broke off, seem to have bounced back and put on a growth spurt generally. A wide range of sizes, with the biggest beginning to head up.

Sweet potatoes: More good news; all plants have recovered from transplant shock and are growing well.

Cucumbers: Not so good. Two of the five sprouts have been (presumably) slug-zapped. At this rate there will still be nothing growing on the trellis after two plantings.

Popcorn: Poor germination; replant in the skipped places.

Beans and sweet corn: Time to thin.

Sprouts: Recovering nicely under Reemay from the baby-rabbit assaults.

Container eggplants: looking much improved, with new leaves which are not speckled.

Eggplants in the row doing fine, too, one growing pretty well.

Potatoes: As noted yesterday; tomorrow I'll finish replacing the straw with excavated soil. The healthiest potatoes we've got are volunteers in the pepper/eggplant and onion beds, particularly the former, which is flowering away and far healthier looking than its poor slug-tattered, mouse-murdered relations in the straw.

Currants: Phenomenal growth, beautiful!

Apples: The Prima has a foliage disease, little raised yellow spots.

Blueberries: nothing ripening yet, though the berries seem full-sized. Query: put up the netting before we leave Saturday for Shawnee? (Query to the query: when will there be time?)

Tour completed, Ted and I set about our next task: installing padding to protect the pond liner from getting torn on cut roots and rocks. We hauled all the old and new carpet remnants we had on hand out to the pond, spent a few minutes shaping and refining the excavation, then proceeded to measure and cut (i.e., slash with an old knife) pieces of brown and blue carpet to fit the bottom, sides, and shelf. The sides are solid clay, but as they dry out a certain amount of crumbling takes place and it seemed prudent to get the soil covered as quickly as possible.

We'd decided to rig a tarp to cover the hole while we're away, but the padding project went so well that I began to consider whether we shouldn't just go ahead and put in the liner and fill the sucker up, so the water could be "curing" in our absence. Ted got alarmed at this— he thought it would take hours and hours, maybe all night—so I called Liz (veteran of two pond diggings and fillings), who said to allow four hours but that we probably wouldn't need that much time.

Ted remained apprehensive but went along. We heaved the liner in its carton—very heavy—into the garden cart and hauled it out to the driveway, where we spread it out to hot up; like the trickle lines, also made of black plastic, it soaks up the sun's heat fast and is easier to work with when warm. Meanwhile I dashed off in the car to buy two lengths of chain and two padlocks to lock the gates, a wise precaution for those with deep pondsful of water in their yards, and a box of four-inch nails to pin the liner temporarily to the soil around the perimeter.

When I got back we dumped the liner, now warmed and flexible, back in the cart and hauled it to the hole. We spread it out on the grass long enough to figure out which way was length and which width, then gathered it up again and draped it loosely over the hole. It fell down inside in great folds; it's at least twenty feet by fifteen—one humongous sheet of black plastic! We stretched out the folds and weighted the edges with rocks and bricks. Then there was nothing left to do but rig the hoses (three of them, connected end to end) and start filling.

Liz was right; it took just under four hours. I found it both exciting and soothing to watch the water run in, and after a while I brought a chair and sat there, the longest time I'd sat down in three days, ostensibly to watch the liner and straighten it out, ease off the bricks, fold the corners, etc., but really just to luxuriate in the sound of water splashing into water. A couple of bees immediately drowned. When the pond was no more than half-filled I saw a flying insect, bigger than a mosquito, dipping its abdomen at the surface time after time, laying eggs the way a mosquito does.

Ted suggested a picnic, brought a chair for himself and went to pick us up a pizza. Time passed; the pond got fuller, the light dimmer. About 8:00 I turned off the water, which by then was about four inches below the rim. For a not-so-level pond, this one's not bad—no more than an inch off all around except in one spot we ought to be able to scrape down a bit before we put on the coping. I was so tired I forgot to bring in the camera and tripod that I'd been using to record these events, and had to go out with a flashlight about 10:00 and retrieve them.

After the pizza I opened the hives briefly to check the supers. The frames in the top super on the swarm hive were full and almost completely capped; the super underneath it, still undrawn four days ago, now has fully drawn comb on nine of the ten frames of foundation. One frame has some honey. I moved that super *above* the full one; that way I can monitor its progress without having to take anything off but covers.

The super on the parent hive has less room left than before, but it doesn't look like the bees need to be given another one before we go away. At least I hope not, because I don't think there's time to build that many frames between now and Saturday.

Obvious in any case that there's a honeyflow on. The bees have been working like little maniacs. Good for them.

June 12. Before doing anything else this morning I finished filling the pond. The black water lapping at the edges looks wonderful, but almost at once I noticed leaves from the black walnuts on the cemetery fence line falling in. Black walnuts produce a substance—juglone—in roots and foliage and presumably bark, toxic to fish; you're warned not to let the leaves fall into the water. I'd forgotten—though Ted hadn't—that the walnut trees drop some leaves every summer. What timing. Later in the day I pulled the bird netting off the strawberries, which we've been picking and eating every day, and which are now just about finished, and recycled it immediately into service as leaf netting above the pond, tied at each of the four corners to the stakes we used to frame our excavation site.

But I've been fretting a lot about going off for a week and leaving the garden to the mercy of any stray woodchuck that happens by, so the main labor of the day was setting up the electric "fence"—a single strand of wire, with lots of special fittings. This involved a number of preliminary steps: unpacking the cartons that have been standing in the living room for six weeks, cutting the fiberglass posts in half, a run to the store to buy a twelve-volt battery, even a call to the company in Iowa with a question about diodes (unnecessary) and the mounting brackets (wrong size; to be returned and exchanged).

Ted mowed close all around the garden fence while I clipped weeds and cut back honeysuckle. We sawed the long rolls of three-foot plastic mulch in half like logs of firewood, spread the resultant strips of flimsy film on the ground along the fence (to discourage weeds), and nailed them down with fiberglass rods. More plastic in my homestead, alas. Then the clips went on, and then the wire on a fat reel, six inches above the plastic. Ted threw some grass clippings on the most flyaway lengths of Miracle Mulch; he's been cutting grass for two days and has a mountain of the stuff. We ran out of light before we could get the solar panel mounted, but it looks like a very simple business to clip energizer to battery and battery to panel, even without brackets or a proper mounting. For one week, just the battery would probably be enough to zap a woodchuck so it knows it's been zapped.

While we were working, the phone rang: Steven Van Gorder, director of Rodale's late, lamented aquaculture experiment, returning yesterday's call from me. He was able to give me the name and number of somebody north of Philadelphia who has a hatchery and sells catfish and bluegills. So it doesn't look like I'll have to send to Mississippi for my catfish after all.

The beans and corn have all been thinned to four plants per pole or hill.

Bill came by to learn the code for the alarm system he'll be turning off and on while we're away; he and Jeff are about to start renovating the bathroom. I showed him the barn-bee holes—two, and a third just started—and was assured they won't weaken the rafters seriously. For the record, the proper name is *carpenter bee*.

June 13. Today, the end of a week of perfectly splendid and splendidly perfect summer weather—cool in the shade, hot but dry in the sunshine—makes me wish we weren't going away. In fact, we didn't leave today as planned; I just couldn't get ready in time. But this terrific push to get ready to be away for a week has brought the garden forward to the point where I feel I'm almost on top of things, with the result that I hate to go! If we were staying, next week could be the sort of time I keep striving toward and never quite achieving.

Of course, if we *hadn't* been planning to leave, this admirable state of affairs would not exist; we'd be halfway through digging the pond, with the electric fence still packed in its cartons.

But it's not. This afternoon we put it all together: battery, energizer, wires, and couplings. With minimal directions and barely sufficient understanding of what I was doing, I stripped wires and attached clamps and insulators. The big clawlike energizer clamps go on the battery terminals and the little solar-panel needle clamps nip onto those clamps, like a child holding hands with a grown-up, and the last needle clamp grips the fence wire in its teeth. I went around to the farthest point, by the gate, to test it: nothing. Ted then noticed a transparent plastic cover on the negative pole of the battery; he removed that and reconnected everything, and this time the wire gave me a powerful pinch of a shock.

That shock came courtesy of the battery; but later Ted unhooked the battery from the energizer to cut the grass beneath it without disconnecting the energizer from the fence, and got a shock purely on the strength of power being collected by the solar panel! I tested it again myself with the same results. We stood there proud as peacocks with our tingling fingers, feeling that this is what high technology ought to be all about. In your face, marauding woodchucks and baby rabbits!

Weeds are a constant menace to such a low-slung line, but it has to be slung that low to shock these low-living marauders, and that means it will constantly need to be checked and cleared; the plastic mulch should help a lot but weeds are weeds.

This morning I removed the rest of the straw from potato Bed One and finished dumping topsoil from the pond excavation around and onto the plants, such of them as remain. One of the biggest either got run over by the cart or chomped by the rabbit or both; it had flopped over onto the grass and had several broken tops. Only about two Rote Erstlings left. This method of growing potatoes has involved more work for less results than any other way I ever did it. I used up my whole pile of topsoil on Bed One; for Bed Two, I pulled the straw off the back, then just dug soil up out of the empty part of the bed and hilled up the two plants growing back there. The rest will have to wait till we get home from Shawnee.

These bloody walnut leaves are the limit. They are (1) toxic, (2) falling before summer has even officially *begun*, and (3) so little that they go right through the bird netting. Raising the netting above water level just means they fall through *and* blow under the edges. Somebody will have to skim the surface every day. First item to purchase when we get back: a leaf skimmer.

Some of last fall's vetch seed washed out of the beds and into the tangle at the base of the garden fence, where it germinated, grew, and is now flowering very attractively in purple clusters. I enjoy all these hedgerow effects a lot, and hated having to hack all that honeysuckle down yesterday so the fence could be installed.

I put on my bee suit and veil after tea and cut the grass around the hives, which had gotten very long. After this dry bright week, I'm also watering everything prior to going away. It all looks so pretty and feels under so much better control, that I'll end as I began: it seems a pity to leave it. Left to my own devices I'm not sure I would. But Ted does want to go, and wants me to go with him, and his heroic labors this week have earned him the right to insist.

June 20. In light of subsequent events, the purchase of a leaf skimmer seems of such minor importance as not to be worth a second thought. (Not really, but that's how it feels right now.) We returned from Shawnee to discover that a Blessed Event had occurred in our absence: the successful, but one month premature, delivery of seven little black Cayuga ducklings.

Obviously there was some screwup at the hatchery. The woman I spoke to there, a couple of weeks ago, had said she would send me a note to let me know whether I could still order Cayugas after July 17. No note ever came, but the ducklings have arrived—seven of them, not six, the extra obviously to ensure live arrival of *at least* six.

By mid-July I would have had all the right equipment and feed on hand. Now I'm *totally* unprepared, except for my two books (and a Garden Way booklet) on raising ducks in your backyard. Lifesavers, these books.

When Ted jumped back in the car with our boxful of mail from the post office, announcing "Better get home right away—we've got ducks!" it felt like a genuine twenty-four-karat dyed-in-the-wool catastrophe. At Shawnee I'd worked out what ought to have been a manageable schedule for doing all the absolutely essential things between Saturday and next Thursday, when I leave for four days in Michigan: one day to set up the blueberry netting system, one to construct the screenhouse, and plant melons in it, one to plant the pond. I'd ordered the pond plants from Lilypons while we were at Shawnee, specifying delivery between Saturday and Tuesday. (They haven't come yet, thank God!) Now these plans, which had given me some sense of control over things, have been knocked into a cocked hat. It took me twenty-four hours to stop reeling from the blow.

The post office had attempted delivery on Tuesday. Neighbors told the carrier we weren't home. In a very neighborly spirit—this must have been convenient for nobody—the three families nearest us consulted together. One has a big dog and one a cat; the third family is the beekeeping Castellans, and they took the ducklings in—put them in a cardboard box lined with newspapers, while the cat-owning neighbor, Jan Purbrick, drove out to McCullough's Feed Store in Gradyville and bought four pounds of duckling crumble.

When we got home, neither Jan nor any of the Castellans were home and I had no idea where the ducklings were, only that "a neighbor" had them. From the third neighbor, Betsy Stedje, I got the story; but Betsy didn't know which of the other two was presently keeping the babies—the Castellans were going away for the weekend, she thought they might already have left. Not knowing where the ducklings were was almost worse than having them arrive a month too soon. While waiting agitatedly for *someone* to get home I kept trying to call the hatchery, but it was late Friday afternoon and nobody would answer the phone. When neither neighbor had returned after several hours I went back to Betsy, who has a key to the Castellan's house, and asked her to let me in. Betsy and her little boy, John, and I all trooped over. While Betsy was unlocking the front door, John was peering through the window. "The box is still there!" he announced. Indeed. The first thing we heard when we got inside was a chorus of pitiful peeping. And there they were: seven balls of fluff, huddled together. They'd spilled all their water long since; the pie pan was dry and the newspaper soaked on the bottom of the carton. It was cold in the house, probably 65° or so—very cold for baby ducklings.

I picked up the box and Betsy grabbed the bag of crumble, and we carried them home. As soon as they'd been given some water and feed—which they gobbled desperately, poor little things—I fetched out the duck books and skimmed frantically for duckling basics. These babies had been in a box in a cool house, on newspapers, for three days. They'd probably been handled some by delighted and fascinated Stedje, Purbrick, Haselberger, and Castellan children—I wouldn't have been able to resist them if I were a kid. Moreover they'd been splashing around (and pooping) in their drinking water. In a normal summer this wouldn't be so bad; but this isn't a normal summer, and wet duck down has no insulating value whatever, in sleeping bags or jackets or on the actual duck.

My books said "the first twenty-four hours after the ducklings arrive are *critical.*" They must have food, water, and rest, their bills should be dipped in warm water, they shouldn't be handled or disturbed. The books said newspapers are too slick for ducklings and can cause them to develop spraddled legs, that it's of the utmost importance to keep them dry and warm—at a temperature of *90 degrees* for the first week, to be lowered by five degrees each week thereafter—and to prevent them from walking in (and dirtying) their water and feed.

Chicks and ducklings are one day old when mailed from hatcheries. The postmark on the shipping carton said June 15—Monday. This meant that I had on my hands a batch of five-day-old baby ducks which had *never* experienced optimal conditions. How serious this early bad treatment may prove to have been I've no idea. The book said if they were noisy and huddled together, it meant they were cold. My little ducks clumped together and piped continually. Clearly, the next necessity was to get them dry and warm.

I found a clean cardboard box, taped the flaps upright to make it twice as deep, and put in a couple of inches of straw. Then I put the ducklings in, one by one, trying not to squeeze.

Somewhere I've got a light clamp, once used to hold a grow light for seedlings, but a quick search of the house failed to turn it up. Then I thought of the reading lamp I used to have on my exercise bike; it too has a clamp and takes a 40-watt bulb, enough to generate a significant amount of heat. So I got that little light and clamped it to the side of the box, and covered the box with a towel. And presently things got quiet in Duckville. Score one.

I slept lightly and woke early, worried that the babies might not be okay. (By this time, obviously, I'd accepted my fate and given up my attempts to adopt them out. They're absolutely adorable, and heck, they're *here*.) And indeed they were not 100 percent okay. They had drunk or spilled all their water, and for a switch they were too *hot*—it was 94° in the box by Ted's thermometer, and the ducklings were all huddled as far away from the lamp as they could get. Adjustments to the towel took care of that.

This morning I opened the bottle of Bee Go that had arrived just before we left for Shawnee. It was packed in lovely clean wood shavings, and my first thought was: what perfect litter. I scrounged up a new and bigger box from the basement—a poultry box, appropriately enough, waxed inside and out—and dumped in the shavings, then transferred the little ducks. They immediately tried to eat the shavings, picking them up in their beaks and jerking their heads to try to make them go down, but they're still too little. By a couple of days from now, when they might have a chance of succeeding, they'll be back on straw and these nice manure-soaked shavings will be in the compost bin.

In the course of this day I've devised some improvements to the basic food and water arrangements. The pie pan, with a soup bowl inverted in the middle, makes a round, narrow trough for feed, and a small bowl with a jar of water inverted in that makes an even-less-vulnerable waterer, one they can't get into, soak their down, and get chilled (though they still do manage to poop in both dishes). They grab a snootful of duckling crumble, then run to the water dish, dip their beaks and tilt their heads back, and the feed goes down (but not the shavings, which just fall out the side). I've also given them a little dish of sand, for grit.

I swear they've grown visibly since yesterday.

In spite of all this duckling distraction, today we got the netting rigged over the blueberry hedge—probably our most disagreeable seasonal task, owing to the vexatiousness of working with bird netting, which is practically invisible and snags on *everything*. The system consists of five sections of fencing ten feet long by two across, five big pieces of bird netting, enough one-by-twos to weight the netting securely at ground level, and some garden stakes, mostly rotted short. The bushes can spread out sideways as much as they like in future years, but can't be allowed to gain much height or the system won't fit.

First Ted mows the grass short all around the hedge; it won't be cut again for a month. We start by shaping each piece of fencing into an arch over the hedge. The cut wire ends are poked into the ground on either side, and a stake is woven through each side and hammered into the soil to support the arch firmly. The five arches are spaced to be equidistant, with one at each end of the hedge. When all are set up, the sheets of bird netting are unfolded one by one, stretched between arches of fencing, and secured with clothespins. (Ted and I have never yet managed to complete this thoroughly maddening chore without squabbling.) As we work our way along, we lay the one-by-twos on the netting's loose bottom edges to hold them in place. When the whole hedge has been netted, I lace up the ends with twistem wire, and finish off with a couple of strategically placed rocks.

However painstakingly put together, the system has never yet succeeded in excluding 100 percent of the birds and rabbits frantic to get in and ravish the blueberries; birds and rabbits both are demonically clever at getting through fencing of every kind. It does enable us to get a crop. I frankly doubt that the crop we get is worth all this effort and feel generally that the blueberries haven't proved to be all that successful a part of our edible landscaping, but they are part of the homestead now and as such require—or seem to require—that we do all we can to reap what we have sown. But I'd far rather be put to all this trouble for currants or gooseberries.

Anyway, that's that job done for one more season.

I opened both hives briefly. The bees in the swarm/divide hive are capping their second super, which is nearly full. We'll get two supers of honey for sure—though if I hadn't caught that first swarm, I wonder if there'd be *any* surplus honey this year.

We found the garden in good shape, reconfigured potatoes looking wonderful, tomato plants enormous (but no little green marble-sized fruits—the nights have been too cold for fruit set), cabbages beginning to head up in earnest (and still no sign of worm damage). The corn should have no trouble at all being knee high by the Fourth of July, and the beans are all vigorously climbing their poles. We now have *one* (and a half?) surviving Jazzer cucumber(s). The onions are starting to fall over, the sign that they're nearly ready to be harvested—the necks of some are big as leeks. Sweet potatoes are slug-chewed but vigorous. Reseeded sweet corn and popcorn are well sprouted. Flea beetles have hit the in-ground eggplants. Last night's slug patrol annihilated many dozens of fat orange-clay-earth-colored slugs from the potatoes and cabbages. In the rain gauge: 0.9 inches.

A few weeds were touching the electric wire, which makes little clicking sounds at each point of contact; the wire also sags at one place and touches the plastic mulch. I pulled off one violet leaf and got a shock—but you can do it safely between clicks if you're fast. Obviously the fence still works even if shorted out by contact points, though the charge delivered would be less. Very nice to feel the fruits of all our garden labors are safe from woodchucks, as they surely must be.

The pond appears to be on schedule with its algae bloom. The number of walnut leaves on the bottom or soaking in the netting at the surface doesn't appear to be great—maybe this is a nonproblem. Now was supposed to be the pond's big moment, but it's been rudely shoved aside by the duckling emergency. The arrival of the aquatic plants should put it back in center focus though.

June 21. This morning Ted mowed the strawberries to smithereens in preparation for putting up the screenhouse, and when he was done, that was the absolute end of five years' happy harvesting of the strawberries planted by the former owners. Strawberries normally get completely replaced at least every other year, and there was so much foliage disease in the plot after this much time that I didn't feel too bad. But a little bad all the same, and worse about some asparagus plants they'd stuck in among the berries, which were perfectly healthy.

The screenhouse is my attempt to raise melons in an environment from which striped cucumber beetles have been excluded. Thanks to cuke beetles, which carry bacterial wilt, my luck with vine crops generally has been terrible. This year's attempt to direct-seed Amber melons and Jazzer cukes—each reputed to be wilt and/or beetle resistant—in the trellis bed hasn't worked out; the melon seeds rotted and slugs got all but two of the cucumber seedlings. Fine; now we try growing dwarf cantaloupes inside a house made of screening. I'll have to hand-pollinate, and probably won't get a crop anyway; it's really late to be starting vines from seed. But at least I should find out whether they do or don't die of wilt.

I slogged out with my shovel and finicked around with the various pieces of framed door and window screening we inherited with this house, deciding finally on two screened doors for the sides, a patio door screen for the roof, and no clear idea how to close up the ends. Having settled (more or less) on my structural elements meant that the dimensions of the screenhouse were also settled. I decided on its footprint, square to the north/south beds in the garden, and started digging. It was necessary to jump up and down on the shovel every time; the ground was packed solid, exactly like the top layer of soil we removed while digging the fishpond—which is only a few feet away, on the other side of the boxwood and the one azalea. I'd considered double-digging this little plot, but changed my mind when I saw how hard the single dig was going to be.

I don't really weigh enough for this kind of work, especially where I hit the old, dense asparagus crowns. These came up looking like something from the bottom of the sea (or another planet): ugly as sin and almost impossible to get the shovel into. I really hated yanking out these unoffending crowns, but couldn't get them out without damage and really had no place to replant them. They should never have been put in the middle of the strawberries—somebody's idea for a perennial bed, no doubt, but a bad one, since the spears always came up right when it was time to put bird netting over the strawberries and would grow right through the netting. The strawberry crowns came up easily, those that were left.

When I'd finally, with much effort, finished digging the bed—five by seven feet or thereabouts—the next step was to dump on the last of the overwintered compost and rake it in. Since I won't be able to get in and weed, once I've bug-proofed the thing, I covered the bed with pieces of Miracle Mulch, brown plastic with pores, the same stuff we used to make a weed-free base for the electric fence. *More* plastic. Then I went back inside to consider my motley collection of bush cantaloupe seeds, finally deciding to plant all of them—five different varieties in each hill—and see what came up.

I slashed six equidistant X's in the plastic, much as I did for the sweet corn, and mixed a small handful of Vermont Organic Fertilizer into the soil. When the seed was planted, Ted and I partially assembled the screenhouse around it by driving short stakes into the ground on either side of the side panels to create slots to seat them in. We did a certain amount of pulling up and redriving, but in the end the roof sat firmly enough on top without attachment, despite a stiff breeze that was making the bee barrier flap and fill.

It remains to be seen whether lateness and this summer's cool temperatures, in combination with the reduction of sunlight created by the screening, will allow any of these varieties of melon to form, let alone ripen. But seeing what remains to be seen is what all these experiments are about, really.

The Castellans came home while we were working, and after a while Lynn and Liam came over to see the ducks. Lynn was abashed at the light and towel; she'd hoped that since their house isn't air-conditioned, the babies would be warm enough. She also said the kids had thoroughly enjoyed (and manhandled) them. It looks to me as if all seven are thriving, so they couldn't have suffered much damage from the rough conditions. Certainly they're bigger every day.

I'd stirred up the shavings with a stick this morning to bring clean, dry ones to the surface and bury the smutty ones, but the smell from the box was so strong tonight that I decided to dump it out and put in some clean straw. The box dripped and reeked; the newspaper underneath was soaked. I carried it all out to the compost bin in the dark, then dubiously put in the straw and the cleaned-up, refilled water and food dishes. Hard to say whether what's dripping out is mostly spilled water or mostly watery poop. Either way, these ducklings need different housing.

June 22. The babies get bigger and bigger, consuming gallons of water, and are almost tall enough to threaten to get out of their carton. Clearly we haven't much time to ponder. We've pretty much decided to build them a box on foot-high legs, three feet by four, with a hardware-cloth floor over a plastic-draped frame filled with straw or crumpled papers or whatever, and set this up in the basement where the smell won't be such a problem. A freezer in the living room is one thing; a poultry farm is another, as even I am forced to admit. But, even apart from the problems of water source and mouse-proof feed storage, it's just too cold at night to put them out in the garage; and so far we haven't even been able to take them outside in the sunshine, as we certainly could have during the first days of summer in a normal year. The basement is the answer.

These last couple of nights, temperatures have sunk into the forties. No little tomatoes to be seen on any of the lush and vigorous plants. The eruption of Mount Pinatubo is being blamed for this extraordinarily cold spring and summer, here in the greenhouse era. I'd hoped my Homestead Year wouldn't be characterized by any *very* unusual conditions, like drought, terrific heat, perpetual rain. . . . I never even thought of unusually low temperatures, but, those in fact, are what we've got.

The ducklings are quiet till they hear us talking, which stimulates a chorus of peeps.

I rinsed the pink Thiram off the Marketmore 86 cucumbers from Stokes—never again!—and planted the whole packet in the trellis bed. Easy enough to thin later; but the problem so far has been reluctance to germinate and vulnerability to slugs. Try overkill and see what happens. The other type, gynoecious and very determinate, I'll presprout and put out somewhere to see what happens. I also stuck a few zucchini seeds into the leaf-heap "potato" beds, where they may do no better than the spuds did—but there's space there and nowhere else, so once again: let's see what happens.

The greenhouse, by the way, is a dead idea. I love my plans for it, and suffer pangs of regret when I think about giving it up, but I must give up *something* or be worked to death, and the greenhouse is the project sacrificed. Some other year.

June 23. At 9:00 this morning, Ted and I sallied forth to build the ducks a brooder box, one that can be converted later on to an outside duck house. It was a beautiful morning (but cool). We set up the sawhorses and tools in the driveway by the garage doors. Ted concurred with my concept of how to build the box, and we set to work with one accord. The first step was making a frame, 34 inches wide and 46 inches long, for the bottom out of one-by-twos salvaged from the former strawberry-netting frame. Over this we fastened a 3- by 4-foot piece of hardware cloth, by bending it with pliers all around the edge and tacking it to the frame. That will be the floor. We used more scrap from the strawberry frame to cut two four-foot legs and two three-foot ones, and nailed them to the frame so it was standing on foot-high legs; and at that point Bill and Jeff arrived to work on the bathroom renovation.

At the same point I also suddenly realized that the box on its legs was not going to go through the basement door. Ted and I smote our brows. Exactly the same lack of forethought when we bought our freezer is why we have a freezer in the living room; you'd think nobody would make *that* type of mistake twice. We considered taking off two of the legs, moving everything into the basement, and finishing building it down there; but that would have meant knocking the box apart when the time came to move the ducks outside. Luckily, Bill thought of a remedy: pull out the nails and fasten the legs and panels together with screws. That way the pieces can be taken apart and reassembled easily and quickly.

So that was what we did. Or rather, what Bill did; he and Jeff pitched into this project without being asked and saved our bacon for us. I couldn't begin to guess how many times they've done that by now.

We met Bill by a kind of fluke. Before we committed to buy this derelict house, our realtor, Elaine, brought over a contractor to look around and give us some idea of how much the absolutely essential repairs and renovations were going to cost. It happened that this contractor was Elaine's brother-in-law; but what might have been viewed as a case of blatant nepotism turned out to be one of the nicest things anybody ever did for us.

After checking things out and poking around, Bill quoted us a price much lower than expected and was hired. By the time the kitchen and extension repairs had been completed and the locust trees cut down, the roof had revealed a leaky nature. It went on like that with the result that during our first six months in this house, Bill and his son Jeff were more-or-less constant presences in our lives. And over the years they've been recurrent presences ever since.

It's hard to imagine two pleasanter people to have around the house. And however glad I've sometimes been to see the pair of them pack up their battered old white truck for the last time, at the end of a long, disruptive job, the sight of this same truck pulling into the driveway when the next job is to begin invariably brightens my spirits and sends me flying to the door.

Fathers and sons don't often work together as a team nowadays, which says a lot about Bill and a lot about Jeff. Jeff is shy, quiet, powerfully built; talkative, maverick-minded Bill is the opposite, with a stock of fablelike anecdotes to illustrate every situation that arises. (I think we've heard them all, and more than once. God only knows how many times Jeff's heard them.) A former diver, smaller than his son, Bill is graceful and economical in the way he moves. Not only in the way he moves, either. The Wasches have saved us thousands of dollars by underpricing the going rates in *everything*: electrical, plumbing, carpentry, painting, roofing, concrete, hot tar, tree work. They do it all. Not to mention providing free lessons in the sorts of handy skills that, in the Fifties, weren't considered important for girls to know.

Bill's the kind of guy people mean when they call somebody "the salt of the earth." He's also what they mean by a "character." His ancient van spends half its life in the shop, but Bill never thinks of replacing it ("I *love* my truck," he says)—maybe because it wouldn't feel right to mix up a load of concrete right in the bed of a new one. Generally speaking—as his truck-loyalty demonstrates—he and I are in perfect agreement that repairing a thing is better than replacing it. This belief, which makes Bill a true rara avis among contractors, also qualifies him as the perfect *homestead* contractor (if the term's not an oxymoron). The person most impacted by the downside of Bill's thrifty ways is his wife, Grace, who told me with a kind of affectionate despair that, years after Bill and Jeff had redone this kitchen, *our* battered electric stove was still in *their* basement (along with tons of other pack-rat hoardings). She also confirmed my suspicion that Bill would sooner get in his truck and drive twenty miles than pick up the phone.

This phone phobia has sometimes made for difficulties, particularly in view of the truck's off-again, on-again working condition. ("I always *mean* to call!" he told me recently when I leaned on him a little about it. "I meant to all day long. I just don't know why I have such a hard time doing it.")

But life with Bill is a trade-off that works. For putting up with what has sometimes seemed interminably-dragged-out projects, and with uncertainty about whether he'll show up on a given day, we get solid, honest work, utter trustworthiness, and all sorts of extras small and large, thrown in for friendship's sake and for free—along with a good-hearted willingness to pitch in and lend a hand at moments like this one, when Ted and I have gotten in over our heads. I think he positively *enjoys* pitching in, showing us how to do the job right (and in the process doing much of the work himself). A lot of people could testify to Bill's generosity with his time and effort—one reason he sometimes doesn't show up when expected, and why the "official" job at hand often proceeds so slowly.

This time the official job was the bathroom, but the morning was spent mostly on the unofficial duck brooder. Bill and Jeff found no fault with my design, but knew much better than I did how best to execute it; instead of fumbling around like we were doing, all their moves were practiced and skillful. Miraculously, the pieces of scrap plywood we were using for the sides—one of which came from the crate in which our effects had been shipped back from England in 1986, six years ago—were close to being exactly the right width. Ted used a hand saw where he could, and a table saw when he had to, and in no time the panels had been cut, fitted, and screwed to the legs.

By this time it was nearly noon, and I had to leave at 1:00 to go into town. I dashed in to take a shower; Ted and the Wasches took the box apart and carried the pieces down to the basement. Then Bill and Jeff finally got to work on the bathroom while Ted and I put the box back together. I'd marked the side panels L and R for easy reassembly before taking out the screws, but forgot to mark the bottom F(ront) and B(ack), and we guessed wrong. The final product, all screwed back together, had a little gap in one side.

I minded this only because the parts had been carefully fitted to avoid exposing the ducklings to any sharp wires. But this isn't after all a precision job, just a rough-and-ready solution to the urgent problem of animals in our living quarters. Said animals were upstairs by themselves and carrying on to beat the band; they'd eaten all their crumble and drunk or spilled almost all their water (and what was left was filthy), and their straw from yesterday was soaked and packed solid on top with manure. I couldn't *wait* to get them out of that mess and into their sanitary new quarters. I raced outside to get a plastic sheet (for the basement floor) and an armload of straw to put on the sheet, under the mesh floor of the box on stilts, to soak up drips; and while I was seeing to that, Ted put the clamorous ducklings into a box and brought them downstairs.

When we first turned the babies loose in what must have seemed a cavernous space, they ran wildly from one end to the other, peeping loudly. Ted thought they liked having all that room to race around in, but to me it looked more like panic. I cleaned up their food and water dishes (both crusted with guck), before filling them and placing them in the box. The ducklings went crazy, gobbling and drinking, rioting around from one end of the strange floor to the other. They must have been completely empty; for the five minutes or so I had left to watch them in, not one pooped a single time. They never go that long.

We rigged up the new clamp and put in a hundred-watt bulb, but this situation, for all its advantages, is a lot less cozy than the old one. There's no way we can get the temperature of a raised box with a wire mesh floor up to 90° or even 80°. It's just as well that these are such tough little ducks.

While I was gone, Ted figured out a way to lower the light bulb, and used a cardboard partition and a cardboard lid to create a warmer sub box about half the size of the big one. He also hung a thermometer in there—an excellent idea—and added the reading lamp to the heat-producing power of the new clamp light. Despite all this, when I got home it was still only about 68° in there—much too cool, according to my books. On the other hand, the ducklings were as far as they could get from the warmest spot, sleeping all together in a big heap of down—*brown* down, I suddenly realized in this stronger light: the little black babies have turned brown!

Obviously they needed more water than they were getting from the little bowl with the inverted jar in the center. I got an empty wide-mouth peanut-butter jar out of the recycling bin, filled it, and inverted it in a cereal bowl, and set it in the box. Such a rush I never saw! The water in this bowl is deep enough that they can dip their entire beaks and part of their heads in, and with enormous excitement they proceeded to do just that, peeping and throwing water around with wild abandon. No question that this head dabbling is something they need to be able to do. They've begun preening and standing on tiptoe to flap their stubby wings. Growing up fast! They're also pooping again at the same rate as before. Most of their excrement is liquid, but not all the solids pass through the mesh, meaning that the manure problem isn't completely solved.

Worn out, we went out for fajitas and chicken. When we got back I went down to refill the water jar; but the ducklings were resting so peacefully and quietly that I decided to leave well enough alone and crept away.

A tour of the yard and garden left me with a powerful impression of messes incompletely cleaned up and jobs left partly done. Maybe in a way it's just as well that the Lilypons order, timed to arrive between last Saturday and today, hasn't shown up yet; but I'd better call them first thing in the morning.

June 24. The call to Lilypons was disappointing; I guess they're not the classy outfit I took them for. The customer-service person listened to my story, then connected me to her manager, who kept repeating over and over that the plants were guaranteed for a year, and if anything died, they would cheerfully replace it. I replied that I didn't want to invest time and energy into a questionable situation which was not of my own making. The manager agreed that the fault was entirely theirs, but that they weren't going to reship the order since there was "a possibility" that the plants would be fine. I asked, was this possibility a probability? She gave it a 90 percent probability rating. I could get no further with this, so for the second time in a week—the ducks were the first—I've got to take up the slack for a mistake made by a mail-order company.

Had Lilypons told me they couldn't guarantee that the stuff would arrive during the four-day period I specified, I would have ordered it early this week from Paradise instead and saved some money. At least Paradise *said* they couldn't guarantee delivery. Lilypons said "No problem." Nuts!

We drove out this morning to McCullough's feed store, where

Jan Purbrick got the duckling crumble, and bought a plastic waterer for what may be the best three dollars I've spent this year. I've had no more than five or six hours' sleep each night since we got back—worrying (with good reason) about whether the babies had water, then being unable to get back to sleep after tilting the jar and filling the bowl.

Something needed to be done, and the automatic waterer is that something: a plastic gallon jug, set on a base tray grooved to let water flow into the tray until the level rises to the rim and cuts off the air in the groove. When the ducklings drink and lower the level, air bubbles in and water flows out. Perfect. They were suspicious when I first put the thing in—the base is red, it's much larger than the jars and bowls—but quickly discovered there was water in there and started throwing it around and dashing wildly up and down.

They seem *much* bigger and . . . older now. Despite the temperature of (finally) 72°, they appear to be perfectly comfortable. They rest in a little group under the light, but don't huddle together or peep. The book says that Cayugas are one of the hardiest and quietest breeds. They're also hitting the grit pretty hard now, whereas before they pretty much ignored it (except to poop in the dish). And it only takes them five hours to drink or splash out nearly the whole gallon of water.

I cleaned their food dish up this morning, and I think once a day will do from here on out, now that they're such big kids.

In addition to the waterer, we also bought some pelletized feed to try them on in a week or so. What I'm really looking forward to, though, is turning them loose in the garden.

I went out in the drizzle and planted some of those bush cukes I presprouted between brussels sprout plants; they were germinating— little white points poking out of the small ends of the seeds—and wouldn't have survived otherwise. So many fat orange slugs were on the marigolds and volunteer potatoes, and the mulched ground around them, that I went back for tongs and a pail of soapy water and did a speedy slug patrol. Marigolds are supposed to *repel* pests! The slugs obviously love them. Let's hope it won't be long till the ducklings can start converting slugs into duck meat.

Later: just went down to see if they still had water. They do—about an inch—but have cleaned out their feed and fallen asleep in three little separated clumps of three, two, and two. A more contented bunch of little birds I never saw. I don't believe the low temp requires another anxious thought.

Later still: moving the waterer after filling it before turning in, I accidentally set it down on one little duckling's foot. The baby shrieked loudly and repeatedly and then, when released, went and huddled in a corner of the box, face inward, hyperventilating. Its first experience of pain, obviously, and a great shock. My heart was in my throat—was it hurt seriously, or were its feelings just wounded? If hurt, what could I do? But presently it crept over to the waterer and drank a lot of water, and after that it stood up and started to run around, apparently okay.

June 25. I'm off to Michigan this morning, to investigate one retirement possibility Ted and I have come up with. Friends of ours own some farmland southwest of Ann Arbor. The original farm has been in their family for thirty years or so. Nobody actually *farms* the farm; they all just live on it in three old farmhouses, each beautifully rebuilt and modernized by one brother, Bob Kellum. A fourth house on the property, along with eighteen acres, may be coming up for sale soon, and Ted and I have been considering the possibility of buying it and moving up there. I think about those eighteen acres and visions of sheep and a milk cow dance in my head. But there are complications, and I'm going up to try to see whether or not it seems like the complications could be worked around.

Before leaving for the airport I picked the first modest batch of blueberries, which Ted instantly made into jam. I keep telling him, it says on the box that a partial recipe won't work right. It didn't (again), but the jam is an excellent consistency for mixing into yogurt, so no harm done.

The other job completed before leaving was removal of the duckling litter to the new manure pile established in the garden, back by the compost bins. In theory the straw litter can be gathered up in the plastic sheet and carried out to the pile; but we'd been using an old plastic drop cloth of Bill's, left here after a paint job, and it must have had some little rips. I dripped very dirty water all the way up the stairs, lugging that unexpectedly heavy burden up to dump outside. Ted mopped the floor while I went after new straw and a new drop cloth. While I was shoving the straw under their box, the ducklings went nuts and rocketed around; they *hated* that activity underfoot. An instinctive fear of snakes?

Much—maybe most—of the water under the cage was stuff they'd thrown out with their bills, fooling around. No wonder the waterer gets emptied so fast.

June 29. I returned from Michigan yesterday to find Ted at the end of his duck-sitting rope. He'd changed their litter three times single-handedly—experimenting with crumpled newspapers, there being so much water and the straw so much less absorbent than expected—and worn himself out running up and down to feed and water them. In four days they've grown enormously; necks which were just beginning to show some curvature last Thursday now make a snakelike S-shape, and they must be eight or nine inches long. At two weeks! They need to be got out of the basement pronto, but it wasn't possible to deal with this today. Today was the day we absolutely had to get the plants from Lilypons, which arrived during my absence, into the pond.

And we did; but first we brought out a new roll of chicken wire and made a little duckling playpen in the grass near the pond, where we could keep an eye on things while we worked. We did this by unrolling part of the roll, forming the free part into a circle, and anchoring the circle with three stakes. Then we put the babies in a box and brought them out for their first experience of The World.

They *loved* it—once they'd recovered from the trauma of being carried upstairs, and set about nibbling grass and drilling in the spilled-water-softened earth as if that were the most natural thing in the world, which of course it is. They were delightful to watch; we carried out chairs and ate lunch by the playpen. Better than TV.

When the shadow of the oak tree withdrew, leaving the pen in the sun, we noticed that the ducklings were panting and trying to get into the meager shade thrown by the roll of wire. So we leaned a piece of plywood against a carton to make a little lean-to, where they very happily spent the afternoon, returning to it even later on when the sun was low. I guess they liked the feeling of coziness.

Planting the pond was a pretty straightforward job, made much easier by Ted's willingness to help, and by his keeping me company, which was as much help as anything. I'd talked to Liz last night and learned that in her opinion it's unnecessary to put sand or gravel on top of the soil in pond planters to hold it in, provided we use the heaviest clay we had. The heaviest clay we had lay all around us in great yellow heaps—no problem finding plenty of that. For containers I was using mostly brown plastic dishpans, eleven of them, ordered from Lilypons in case I couldn't find that many brown ones locally—brown so they'd be invisible in the water.

We evolved a routine. Ted would shovel some clay into a pan; I'd stick in the tuber, roots, or stems and scoop in more clay with a trowel. Then we'd heave the pan to the side of the pond, drench the soil with water, and let it sit till air had stopped bubbling out. Then, when several of these were ready, I'd get into the water and place the pans where we wanted them to go.

courtesy of Lilypons Water Gardens

I made the mistake of releasing the snails—whopping big things, plus a few tiny babies—before installing the plants, and had to be constantly careful not to step on or set a pan on top of one. None of the snails showed any signs of life, even several hours after I'd released them from their plastic bag, but only one showed signs of death (by floating). That one proved to have a hole in its shell, and I pitched it into the jungle; the others, Liz says, are probably okay but it may be several weeks before we see any activity.

Most of the plants seemed alive, but pretty exhausted after nearly a week out of water. A second shipment that arrived Saturday demonstrated how much healthier and better they—Anacharis in this case, one of the three species of oxygenating plants—would have looked with a briefer transition time, and I got annoyed at Lilypons all over again; but only about ten bunches of Cabomba and the three parrot's feathers were, respectively, more slime than frond and more yellow than green, so only they will be reordered rather than planted provisionally. The final list: three water lilies (an Odorata and two Comanches); three water clovers and three floating hearts; three each of cattails, blue iris, and Sagittaria for the marginal shelf; Cabomba, Anacharis, and Myriophyllum, 75 bunches of each type. And the three parrot's feathers.

The water was beautifully "cured"—green, murky, opaque—and the liner slippery; I nearly fell once, trying to avoid treading on a snail. There are trillions of mosquito larvae at the surface, also several water striders of totally mysterious origin—Ted says they've been there from the very first—and a water beetle that probably rode in on one of the plants.

Walnut leaves are still drifting down at the rate of a dozen or so a day. The bird netting was of questionable usefulness, so we removed it. I'm simply going to have to hope the fish can tolerate it if I minimize the impact by skimming the surface every day.

The bog plants on the shelf—iris, cattails, Sagittaria—are tilted, won't stand up straight even in this heavy clay mud. I'm thinking I'll get some coarse sand—the ducks will need sand for grit, any-way—and top off the pots, to see if that will support them better. Otherwise, I guess the pond's planted. Now all it needs are fish and a frame.

And I'm wondering, now that I see how soft and disturb-able this mud is, whether it mightn't be better to shift over completely from catfish to bluegills, which aren't bottom feeders or mud grubbers and should leave the soil in the pans where it belongs. Something to discuss with the fishery folk when Liz and I cruise up there on Friday.

This morning, before starting all this, I did the first major blueberry pick. Crawling around under the netting was as aggravating as ever, and it took me an hour to clear all the ripe and almost-ripe berries off two bushes—they're very small berries—but I ended up with over a pound of them in my bowl. Far as I can tell, no intruders have gotten inside the netting this season (so far).

The results of my trip to the farm, by the way, are inconclusive, but I'm not optimistic. Ted's even more doubtful than I am that we outsiders could ever be comfortably at home in a setting so filled up with another family's history and life. Farewell, I guess, to the fantasy of sheep and cows, and back to the drawing board, if nothing happens to change the outlook.

June 30. By their fruits ye shall know them. Counting in from the turf path: in Bed One, tomato plant #4 has revealed her identity as a Gurney Girl, #5-7 are Viva Italias, and #8—well, I knew this—is a Gurney Girl, as are #9 and #10 (still pretty small). Viva Italia #5 is badly hit by septoria leaf spot; the rest have light cases, normal for this point in the season. Except for the mature Gurney Girl in Bed Nine, the vetch bed, *none* of the plants over there has set any fruit. Too *much* nitrogen?? Judging by size, the extra Beefmasters all wound up in Bed Five, the Viva Italias all in Bed One—but we won't know for sure until the plants in Bed Five produce some fruits to know them by.

But (speaking of fruits and recognition) the two volunteer potato plants in the lettuce and eggplant beds blew my mind today by turning up with little clusters of what look for all the world like green cherry tomatoes on stalks. At first I thought a volunteer Sweet Chelsea descendent had intertwined with the eggplant potato, which is so huge and lush it might easily conceal something else within its coils. But no: these little fruits really are growing on potato plants. Liz has documented the phenomenon with her close-up lens. I never saw anything like it before.

The pond still looks like hell, but there are some little signs of progress. The Odorata lily has sorted out its several leaves and turned them flat to the sky; the water clover has made a tiny, perfect mint-green clover leaf and is in the process of making more. Liz says it'll definitely take a while, but things look hopeful. And fish will help. As will trimming the excess liner and getting the pond framed in, and getting or making a skimmer to flick out these infernal walnut leaves (and other flotsam).

Sunday afternoon when I went out to look, the cukes I'd planted in the trellis bed were abundantly up; today, 36 hours later, only one survives. Melon sprouts up in the screenhouse too, but fewer, it seems, than yesterday. Two more slug-vulnerable crops to be started under lights in future, O joy. Better is the news about the presprouted bush cukes I stuck in between brussels sprouts; those are up and looking pretty good—but I'll check in another day or two.

For many hours yesterday I watered the screenhouse and the main and back gardens, taking the opportunity to do some repairs on the trickle lines. Next is the potatoes' turn: Bed One today, Two tomorrow while we're building the duck pen. Unable to find my hose grabber, and needing to overhead-water the potatoes, I set a broken brick on end and stuck the nozzle through one of the holes. Low-tech, but worked fine.

This is the year's longest dry spell, but things have finally warmed up: in the upper eighties or maybe ninety today.

The seven ducks spent another blissful day eating grass, trying to eat acorn pieces, and sleeping in their playpen lean-to. Knowing the day would be hot, I moved the hex-wire circle farther under the oak, and set up the lean-to only because of my impression that they had enjoyed huddling under there, whether or not forced by the hot sun. Sure enough, they spent a lot of the afternoon cozied up together in that lesser shade beneath the greater. Liz photographed them. Bill and Jeff, still here working on the bathroom, brought their beers out under the shade to watch them scamper around.

Sunday these ducks were two weeks old, so as of today I'm reducing their rations to as much as they can clean up in five or ten minutes, three times a day. This is supposed to encourage them to forage more vigorously, and in fact they did spend a lot of time today chomping on grass, that fine sort that grows under shade trees, and also some broad-leaved weeds. Jeff happened to be watching when a lightning bug flew through the mesh and was instantly snapped up by a duckling. Later I saw another land on the wire, and nudged nature by catching it and tossing it into the feed dish. A couple of seconds later, a duckling spotted it. Peck! No more lightning bug. Earlier I'd given them a potato mystery-bug larva. Gone! I'm actually looking forward for a change to the Japanese-beetle assault. Too bad these babies won't be big enough to take advantage of the June-bug hatch as well; but *those* whoppers could eat a duckling in one bite, even a giant two-week-old duckling like ours.

Crop report: the sweet potatoes are very badly chewed, especially certain plants, and have yet to start vining. The onion tops are definitely falling over now; in a day or two I'll knock down the rest. The sweet corn looks wonderful, and the lettuce is *finally* growing like crazy—just in time to bolt, no doubt.

Also, looks like tomorrow will be another blueberry-picking day, as well as a duck pen-building day. Ted made another half-recipe of blueberry jam despite my earlier objections and warnings. The verdict: better than previous batches, but looser than would be optimal (lunch was peanut butter and jam so I speak from experience). I yelled at him a little for this, and he's promised to wait now till we have enough for a whole recipe, and do it properly.

This was not Ted's day. I slept late, and he had wanted to leave early for town. So instead of waiting for me to help, or leaving it for me to do on my own, he tried hauling the water- and litter-filled plastic sheet from under the duck box out of the basement by himself—and had the sheet break on him under the weight of all that filthy water. I got up to find him on hands and knees with a sponge and bucket of soapy water, wiping the liquefied duck poop up off the laundry room floor. Since I'd planned out how I was going to get most of the water into a bucket before trying to lift the sheet, I'm afraid I wasn't very sympathetic.

After tea I tied up the kiwi vines (again). There *are* some tiny green kiwis on the fruiting laterals, but nowhere near as many kiwis as there were flowers. Plus Ted spotted a squirrel munching on something up there this afternoon, so we may not harvest anything.

Liz says the linden trees just bloomed, and that lindens are a major honeyflow in these parts. Oops. Tomorrow I'll check the hives, see if there's room left in the supers or if I need to knock that last super of frames together in a hurry. I'm very grateful to the bees for taking care of themselves during these days of maximum duckling mothering, but certain minimal gestures must be made.

July

July 1. A long, exhausting, but productive day. We disassembled the brooder box, carried the pieces upstairs, hosed them off at high pressure, and reassembled them (with nails, not screws) into a duck house. We also built and roofed the pen (an easier job).

Ted cut a door in the brooder box, then made a roof for it out of some more of the English shipping crate—the best service those highway robbers did us was to build that crate. What had been the reinforced top or bottom of the crate became a perfectly terrific roof, when covered with inherited leftover shingles. The piece of crate was heavy, though, even with half its cross braces removed, and heavier still with the weight of the shingles; so Ted nailed two of those same braces—three-by-three posts, really—to the front legs of the brooder box to carry the thrust of the roof. And to our pleased surprise the house sits there with absolute confidence, its back legs on the concrete apron of the garage and its reinforced front legs in the grass. There's not a wobble, though all measurements were pretty much rough and ready. Typical, actually, of all Ted's building projects: he throws himself into the work, guestimates and improvises, bashes it all together—and nine times out of ten the result is both sturdy and charming.

This one is for sure no exception; it's perfectly adorable, and almost classic in its styling. Ted cut a door in the front panel and made a little ramp with cross pieces for steps. Then we set stakes and stretched the wire to build a pen.

All day the babies had been contentedly ensconced in their playpen, which today was made of twenty-four feet measured and cut off the roll of wire, so the rest of it could be used to build the pen. At one point in the afternoon the threat of a thunderstorm had us working feverishly, trying to get some space enclosed for the downy babies in case it started to rain, because there was now no place to put them inside the house. And it did sprinkle a little; but they stayed put in their lean-to, as happy as seven campers in a pup tent.

But eventually the twenty-four feet of playpen wire was needed to build the roof of the run. So late in the long day, when we were both staggering with fatigue, we caught the ducklings, carried them in the carton to their new house, and stuck them inside one by one, with the feed dish and waterer. Then we blocked the door with a piece of shingle and cut the wire of their playpen into two pieces of ten and fourteen feet. Side by side, these make up the "ceiling" of the pen, without which these little ducks wouldn't survive a single day. I thought they would mind us hammering on the house, attaching the short piece of hex wire with staples and tacks. But they settled right down. Apart from the roof and doorway it's the same box they've been living in for a week, after all.

While we were roofing the pen, Liam Castellan came over to visit the ducklings. He was obviously disappointed at how big they've gotten, and how much less cute, since they lived in a cardboard box in his living room. Claire Purbrick had the same reaction four or five days ago and they're much bigger now. Liam watched us work for a bit, then said, "Well, I guess I'll be going now." But disappointed or not he's agreed to baby-sit the ducklings while we're in Utah.

An aside any DIYer will relate to: this morning we rushed out to the truck to go pick up some materials, mainly a replacement fence rail, and the damn thing wouldn't start: dead battery. I raised the hood and beheld two badly corroded battery poles, which I cleaned up before clamping on the jumper cables. But we still couldn't get it started.

Pause to locate AARP Travel Club membership card, ascertain that it hadn't expired, and place a call. Interval of waiting for the tow truck, during which I hit upon the idea of switching the rotted-out rail in the crucial position with another rail in better shape. As long as there was a sound rail to fasten the duck-run roof to, no need to go to the store this morning. So, thanks to the dead battery, we've still used no boughten materials for the duck shelter except nails and chicken wire. Every time we've needed another piece of scrap lumber of a certain size, one has turned up.

This duck-house project involved a lot of sawing and nailing, and occupied Ted more constantly than me. I did some of both, but as I was leery of putting much stress on my vulnerable right elbow, my contributions were largely in the realms of planning and problem solving. So throughout the day I fitted in other projects as opportunity allowed. I took twenty minutes to open the hives and found no sign of crowding in the supers, so that's one concern off my mind. I *am* puzzled by the fact that in the second hive, many of the capped cells have little perforations in the cappings, as if the bees might be tapping the supplies. Do they do that? I've never read about it, and previous observations suggest that they cover the cured honey smoothly with layers of thin wax, not from the periphery of the cells to the center with thick wax. But no time to research this now.

I also watered the potatoes in Bed Two, using the same broken brick in lieu of the hose grabber, which may come to light when we clean the basement (and that'll be soon, now that the ducks are out of there, and now that they've made the job even more urgent than it was already). The one surviving Marketmore cucumber is still with us, and a tiny true leaf is showing between its big fat seed leaves. Maybe it'll make it. The two Jazzers have gotten too prickly now to interest slugs. For future reference, the surest way to outfox slugs where cukes are concerned is to grow them in a pot till the stems get prickly. I've direct-seeded *much* more successfully in other years, or maybe I'd have reached this conclusion sooner. Some losses in the bush cukes but lots still hanging in, probably thanks to the present extended dry weather.

Three hills in the screenhouse have melon sprouts and three don't. Now I must decide whether to struggle further with this project or cut my losses. I won't go to the cost and trouble of closing the ends till it's clear whether any of the sprouts will survive. The design shuts out insects, but not the resident population of slugs—resident, I mean, in the screenhouse soil.

There's *one* Viva Italia in the vetch bed, fruiting later than those in the control bed.

In the pond, overnight the water clover has produced *lots* of tiny pretty clover leaves. A great choice, that one. Ted thinks the cattails are greener than formerly. The Odorata leaf that was emerging yesterday has unrolled and is lying almost flat on the surface—pretty fast work. We saw a water strider striding across this leaf in a satisfactory way. Tiny signs of new growth in the Sagittaria and floating heart as well.

I spoke today with Richard Kurtz of Kurtz Fisheries about catfish versus bluegills. Bluegills will mess the water up less, he says, and will spawn in a pond as small as ours (fun!) which Steven Van Gorder told me catfish won't do. But bluegills are bony and will compete with their own offspring for food and not grow much. I knew that from my own limited fishing experience, but am inclining toward them anyway if the cats will definitely muddy things up worse.

Liz and I are driving up there Friday, rain or shine, with a five-gallon bucket in which to bring back the plastic bag of fish and water in case—says Richard K.—the catfish barbels poke holes in the bag. Kurtz also sells redworms, so I think I'll get some of those now as well.

July 2. I woke up at six or so this morning, fretting that the ducklings wouldn't be warm enough in their mesh-bottomed house with overnight temperatures in the sixties. When I realized I wasn't going to be able to fall asleep again, I pulled on jeans and a sweatshirt, got the bag of crumble and a bucket of water, and stumped out to tend the livestock.

And made an interesting discovery: that the ducklings make this place feel like a real homestead. Something has been added to the mix that neither garden nor bees nor orchard nor pond, not separately and not together, had brought into being.

Anyway, all was quiet, just as it had been at 3:20 A.M. when I went to listen at the kitchen window. Wouldn't they be noisy if they were uncomfortable, or nervous about spending their first night outside *and* their first night in darkness? Until now they've had the constant presence of at least one light bulb. They were plenty glad to see me—hungry!—though I was surprised to find that they had a little water left. I wanted to try to induce them to come down the ramp, so I put the filled feed dish and waterer down at the foot and waited to see what would happen.

What happened was that they *did* venture forth fairly soon, but each one only went a little way down the ramp before jumping/falling off, thump! on the side where the food dish was. Later I tried to get them to go *up* the ramp to get their lunch, by letting them peck a bit, then showing them that I was lifting the dish into the house. No luck: developmentally they just aren't ready for ramps. They're not chickens, and this is more of a chickenhouse arrangement. I think they'll learn; there's a picture in one of my books of a duck ascending a cleated ramp to get a drink. But not yet.

We needed to put a few finishing touches on the duck house this morning, difficult with the ducks in residence—loose ends of chicken wire flapping around, etc.—and I was dying to see what they'd do if put into the garden anyway, so Ted shooed them toward me and I caught them, popped them in the trusty carton (they *hate* being caught but no longer panic in the carton, though they're now tall enough to see out of it by stretching their very stretchy necks), took them into the garden, and turned them loose. I put the waterer in too. Then Ted brought over two chairs, and we stayed ten minutes or so to watch the show.

They were overwhelmed, that's the only word. So much space! So much stuff! They ate: violet leaves, thistle, marigolds (I kept making them stop), grass, dandelions, and assorted other greens, like toddlers trying things out. They trampled one of my sweet peas, but those are so pitiful I didn't care. They started to eat the peppers and I made them quit. They sampled potatoes and tomatoes but quickly gave that up. Every time they got a little scared, they rushed for the waterer—like a baby mammal going for the teat.

One of them pecked through a thin mulch of grass clippings in one bed, uncovered the trickle line, and jumped—another snake? Yesterday we found a worm under a board, and I tried feeding it to them. ("Don't do that! It's our only worm!" said Ted.) They were scared of it, flinched when it coiled on top of the feeder.

In a little flock, all facing the same way and moving in a peeping clump, they ran through the garden. After a while I decided they weren't going to hurt themselves, or eat anything I couldn't live without, so we shut them in and went off to finish closing the gables of their house. From my workplace at the picnic table I could see them running, all together, back and forth between the rows. They then found the fence and caught sight of me. Peeping loudly, they climbed up the pile of leaves in the corner and stood there, peering over the fence and making a hell of a lot of noise.

"They want their mommy," said Ted. Was that possible? Experimentally I walked over to the fence—and they immediately shut up and nestled down in the leaves.

"Well I'll be damned," I said. Whenever I try to catch them they run; but in a big, unfamiliar space, where they don't feel secure, my presence is reassuring to them. How about that!

But we had to finish the house, so I left them peeping plaintively—they started up again as soon as I walked away—and quickly did what was necessary. One gable we blocked up with a board, the other with cardboard (the former brooder partition) cut to size; till they get older and chilly nights are no longer a concern, cardboard's good enough. Like dopes we forgot that by roofing over the run we were making it impossible to do any work inside it; so the length of roofing hex wire next to the garage had to be detached.

Then we nailed a board across the inside pair of legs, before boxing in the whole thing like a house trailer—again, in the interest of warmth, and also because pushing litter under the house and pulling it out should be easier with a backboard, or so I assume, not having tried it yet. Ted fitted another board across the front legs. The garage wall is the back "board." Not perfectly draft-proof, but like having the windows open a crack instead of wide open.

Then it was time to put the babies back in their pen. They had settled down in the leaves, but were keeping an eye on me and started peeping whenever they couldn't see me. I'd figured we would catch them again and put them back in the carton for the trip, but their dependence on me in strange situations gave me the idea that they might *follow* me home; and Ted, who had read about herding geese with a wand, caught up a bamboo garden stake and went into the garden.

The ducklings ran headlong to meet me, and I waited outside while Ted herded them adroitly through the gate—an amazing sight! Sure enough, they did follow me across the yard, in a general way (I was carrying the waterer and quacking); equally sure enough, it turned out to be perfectly possible to herd them with the stick (and great alertness). I opened the gate of their pen, stuck in the waterer, and in they ran, glad to be back in familiar territory.

Now to figure out how to get them into their house at bedtime.

This morning I did the second, equally maddening and tiring, major blueberry pick. You have to remove all the bottom boards, unclip the clothespins on one side, and crawl completely inside the netting to pick the berries—an awkward, slow, uncomfortable business. One pound six ounces, for a season total (so far) of about three pounds. They'll have to be picked tomorrow, too, because I quit after an hour, having had enough of the physical and mental strain. The berries on each bush are many different degrees of ripeness, I have to crawl around under the netting *making choices*. And they aren't even all *that* good to snack on, even the dead-ripe ones—unlike raspberries. At least the bushes don't have thorns; but in the perfected homestead of the future there will be no blueberries.

But there will be ducks, and a fishpond.

Later: getting the ducklings into their house was a problem I felt incapable of solving, but lows in the mid-fifties are predicted for tonight, and ducklings this age are still supposed to be in a brooder kept at 80°! So what we did instead was get them to go *under* the duck house, where the litter is, and block off all the openings where cold drafts could come in with cardboard, shingles, even the spare hive bottom board. It's very makeshift and desperate looking, but we lured them in with a pan of feed just as it was getting dark, and closed off the front with a board. Then we hung the brooder light inside the house with the bulb touching the mesh floor, making a warm spot underneath for them to huddle in. And that will have to do; but what if it doesn't? Worry, worry. Being a first-time mother is no snap. But they're quiet out there now. I crept out a little while ago in the dark to listen and heard only contented peeps.

July 3. Last night after we'd gone to bed I surprised myself by launching into a Duckling Rave. Between the pond and the trips (I said to Ted), I was overextended already; the untimely arrival of the ducklings makes things feel completely out of hand. I work till 10:30 or so every night and *still* can't get on top of it all, and now the Utah trip is looming and we'll be gone *eight days*—infinitely long in the life of a duckling. I'll miss all that developmental stuff, Liam will have to be paid to baby-sit, the pond will do all sorts of interesting things I won't be here to see, the garden's bound to get wildly out of hand since it's pretty out of hand *now*—all because the hatchery screwed up and sent the ducks themselves, instead of information about them!

If that hadn't happened the pond would be finished and framed in, the garden would be under control, the screenhouse ends would be closed, I'd be planning to extract honey this week, I wouldn't have needed so much help from him—help which interfered with his own scholarly plans for the summer—and, above all, I wouldn't be so strung out. Things would be proceeding as *I* decreed, not as they've been imposed on me by the blunders of others!

I knew as I said all this that I'd feel better about it in the morning, and I do; but not so much better as to deny the essential truth of the Rave. The extra emergency-style work they've created has meant that each morning I hit the ground running, that each job is barely completed in its most essential form (with lots of loose ends left flying, to be done later somehow, not the way I like to work), that our house is chronically a pigsty because we're both too exhausted to put anything away ever, and that I keep dropping stitches, forgetting to do things, forgetting what somebody told me, just generally unable to keep all the umpty-ump balls in the air.

It was chiefly to avoid living like this that I planned the Homestead Year! In that one real sense—halfway through—I have to admit that the year's a failure.

This morning too I was up at 6:30, pulling on my jeans and boots to go do a duckling check. What I found—contented peeps—was reassuring, and when I lifted the board out from its restraining nails they poured enthusiastically out from under the house, looking for breakfast. I'm keeping the hose turned on at the spigot and using the valve at the nozzle end to open and shut off the flow of water, which makes cleaning out the waterer and feeding dish (still the original pie pan with inverted bowl in the center) much easier.

One of the stitches I've dropped was realizing how nearly out of feed we are, and how much faster the ducklings consume it than they did only a couple of days ago. Ted said yesterday that they'd never make it through the Fourth of July weekend on what we had on hand, but I was sure there was plenty. He was right, though. I started planning what they were going to live on till Monday, but called McCullough's just in case and luckily they're open today.

But this has been primarily a fishpond, and not a duckling, day. This morning Liz and I drove up to Richard Kurtz's Fish Hatchery, near Elverson, to pick up the fish for the pond. (Ted, incidentally, went in town for the day, and it was a great comfort to know the ducklings were shut securely into their covered run, beyond the depredations of cats and crows.)

The drive up was a pleasant change from ducklings, and pleasant in itself. Elverson is in that splendidly rolling piedmont country of woods and fields, like Kutztown, like Oley, like the border country of Scotland and Wales for that matter: my favorite type of landscape in the world.

Richard Kurtz turned out to be a very tired young man in cutoffs and those river-bottom-walking slipper-type black "shoes" you see in the catalogs. We followed him into a dark, windowless building filled wall to wall with concrete tanks each about the size of my pond, in which water was circulating so noisily it was hard to talk. I'd worked out that my ± 1,300-gallon pond could accommodate at least 30 four-inch fingerlings, or one inch of fish per 10 gallons of water. But I wanted *plenty* of growing room, and decided, pretty much on the spot, to get only sixteen bluegills and four catfish (just for fun, and not too damaging to the plantings). Just like that, Richard hopped into the tank and started netting bluegills into a five-gallon plastic pail while Liz snapped pictures with a flash. He fished one out with his hands and slapped it on a ruler: just over three inches nose to fin-tip, therefore in the three-to-five-inches category I'd requested. When he'd put sixteen of them in the bucket, he rushed out, leapt onto his minitractor, and roared away to another tank somewhere else—in what was by that time a light rain—to fetch the four catfish.

Then came a final leap into a different tank to net some tadpoles, at least one of which looked enormous. "The big ones will turn into bullfrogs this summer; the others won't do it till next year," said Richard.

He was so rushed and harried—I learned he'd gotten to bed at 1:00 and been up at 4:00 that morning, worse than me!—that I forgot to ask about redworms. Maybe just as well. He offered me some freshwater snails that looked to *my* eye exactly like the Japanese snails from Lilypons, saying they were very prolific. Liz concurred and warned that they'd get out of hand; but then Richard said the fish would eat the tiny ones—and since three of the Lilypons snails have floated to the surface already, I said okay and took six.

The total bill came to $13.78: four catfish at .65 each, sixteen bluegills at .50, six snails at .15, six tadpoles at .15, and .60 for the double plastic bag filled with water and air from a pressure hose, in which they traveled back to Rose Valley, plus tax. Since *one* four-inch catfish from Paradise Gardens, the cheapest of the mail-order outfits and the only one offering catfish at all, cost $4.00 plus handling and air freight, I think I got a fine deal. I didn't count everything, but Richard said he put in extras of "everything but catfish," making the deal still finer.

When I get home with Liz's five-gallon bucket stuffed with a bag of water creatures and lots of air—stuffed so tight I had to lay it on its side to get the bag out—the ducklings were running around in the rain. "What are you guys doing outside?" I said; I'd expected to find them tucked up snug under their house. This morning Ted sawed a foot off the board that closes off the front, to make them a doorway. But downy or not, they seemed impervious to the rain—or, more correctly, they seemed to *like* being out in it.

They were also famished, so—shaking my head—I cleaned up their dishes and stuck the food dish under the house, hoping they'd have the sense to stay under there when they finished eating. Then I heaved the bucket into the dirt cart and trundled it over to the pond. You're supposed to float the bag of fish etc. in your pond until the temperatures of the pond water and the water in the bag are equalized; this minimizes the shock to the fish when they're released. Makes sense to me. I left the bag drifting in the pond like a chambered nautilus, and the ducks scampering wildly back and forth in the rain, and jumped back in the car to drive to the feed store, where I bought a fifty-pound bag of duck pellets and a five-gallon feeder that I hope will work as a waterer Liam can manage—no time to test it today.

It rained like anything all the time I was coming and going on this errand, but the ducks were still out when I got back, running around, shaking themselves and chirping happily. Impossible to believe they were cold and miserable. Obviously nobody ever told them ducklings their size have to be kept at 80°.

I wanted to get a picture or three of the big fish-and-tadpole release, but at 5:45 it was pouring and it seemed better to just go ahead and let them go. So when Ted got home from Penn we put on rain gear and went out together. The nautilus had drifted up among the bog plants. I undid the rubber bands from the outer and inner bags, glanced in (through the rain) at the seething shoal within, then sank the edge and let pond water fill the bag. And out they came and disappeared at once into the murky depths—not to be seen again, probably, till the water clears.

Which should be soon. When we got back to her house yesterday, Liz gave me two water hyacinths for the pond. "Do I want water hyacinths?" I said, aware of how badly they choke the waterways of the South. But Liz says they've been used in experiments to clean up gray water and will clear up *my* water faster than anything else will. You don't have to plant them, they just float on the surface—couldn't be easier. Liz's own pond, by the way, is an inspiration—beautifully clear water, koi, and two generations of lively goldfish. So two little green water hyacinths have also been set adrift upon the surface of my pond (and immediately blew among the bog plants and stuck, but never mind).

It went on raining all evening. I gave the ducklings a generous dinner, thinking they *must* be cold by now, and shoved it well back under their house; but at 9:00 they were still quacking about in the open. When we went out to feed them, and again at dusk, they rushed over for their handout—from the other end of the run, where they'd been paddling happily in the outflow from the downspout that drains the garage roof! *I* was cold—why weren't they? They were starting to be really wet, but didn't have the sense to come in out of the rain. I plugged in the light and we left them there, not knowing what else to do and comforting ourselves with memories of the mallard ducklings on the Cam, who somehow have to get through those cold, wet English spring nights even when they're too big to be brooded by their mother.

All that frantic labor of building the duck house, blocking up the big gaps, hanging the brooder light—apparently not needed. The books must be describing average ducklings, not these tough Cayugas.

July 4. In the rain gauge: 1.4 inches overnight. The triangle of sweet corn has been *flattened* by this rain, reminding me that by now I'd normally have hilled up all the sweet corn and popcorn—raked soil up from both sides of the rows to support the stalks as they get tall and top-heavy. This corn can't be hilled in the usual way because it's planted in plastic. Instead, I'll have to dump dirt on top of the plastic and tuck it around the stalks—a *great* reason not to grow corn in ground so weedy that only plastic will control the weeds. It's lucky we've got those piles of pond-excavation soil—which, by the way, I've still to finish putting on the potatoes.

The last little Marketmore cucumber is gone without a trace. Not so the bush cukes among the sprouts. The sprouts themselves are so desperately slug-chewed I'd despair of them if I didn't remember the cabbages looking just as bad—and we've eaten two big cabbages already as coleslaw, with lots more to come.

Speaking of cabbages: I keep seeing white cabbage butterflies, and little worm droppings on the plants when I check, but never any worms. My theory is that the birds are eating them. There are birds in the garden every evening, perching on the bean poles and tomato stakes, singing, playing in the spray from the hose, doing whatever they do. A catbird, a song sparrow, a cardinal, and a peewee are regulars, and other incidental visitors come, too. I think the peewee may be catching a few bees, but I'm very fond of flycatchers and honored to have one in my garden, even if it costs me a little honey.

When I went out this morning around 8:30, the ducklings, praise be! were under their house. I think they probably went in there at last not because they were cold, but because they were tired and wanted to snuggle up and go to sleep. They came rushing out, and Ted and I herded them over to the garden, where we spent ten minutes taking slugs off the cabbages and putting them on a board where the ducklings could see them and scoff them down. It's not too soon to start training them to earn their keep.

All seven little bills were gucked up good with slug slime in no time flat, so I brought them the waterer, figuring they'd drill in the dirt if they had water to spill around, and the dirt would rub the slime off. (It works on hands.) And sure enough, when we went back to get them after breakfast, their beaks were yellow with clay mud but slime-free.

They're getting faster and harder to herd. Luring appears to be the method of preference. I don't suppose mother ducks ever chase their babies. They just go along quacking, and the babies chase *them*. That's what I remember from watching the ducks on the Cam.

The Prima apple tree has rust in a big way. Much worse, I'm

sure, because I took no action when I first noticed. I brought the ducks a fallen apple with a slug attached, stepping on the apple first to crunch it up, and they gobbled it down. I can't do anything about the tree. It'll have to sink or swim on its own—I don't have time to read up on what steps to take. That's what racing a deadline comes to in this case: abandoning a tree I might rescue, were I able to do certain things.

Ted has made a gate for the pen—very sturdy and handsome. What I would have done without his help in this emergency I can't imagine. He made it clear to me long ago that this project was mine, that he'd be too busy to help much. And all through the spring, that was true; but for the past month he's worked at least as hard as I have at the actual doing, even though all the managerial stuff and record keeping is still my department.

The new "waterer" does *not* hold a supply of water in its central cylinder, but does hold quite a lot in the bottom part and can be filled with a hose from a distance. It's a great success with the ducklings, who are busily rinsing their bills in its deeper trough, doing something else by the book. They spent the afternoon snoozing in the shade of tomato Bed One while we cut a section out of their fence to cover the new gate. Right after Ted left for his Fourth of July cookout they started peeping, so I herded them home—through the new gate—and left them playing in the water while I went off to work in the garden. No rest for the weary, Fourth or not. (One duck had dead leaves stuck to its bill with slug slime. Foraging!)

First thing was the pond. I'd planted the best-looking of the yellowed parrot's feathers in a pan of clay, and stuck the pan in the jungle to await the replacements of the rest. Liz had checked them and pronounced them fine. So after scooching the pots on the shelf together to make space, I filled and sank the pan before pushing the new parrot's feathers into the soft clay with the old ones. I *think* one of the Comanches is sending up a rolled green leaf, but if so it'll be a couple of days before it breaks the surface. The water smells amazingly like a natural freshwater pond; a breeze sent us a lovely whiff this afternoon while we were hanging the gate.

The reason I didn't go to the cookout too was that the flattened sweet corn so urgently needed to be propped up. I filled the dirt cart four times from the mound of excavated topsoil next to the pond and dragged it into the garden, where I shoveled soil around the bases of the cornstalks and pushed it against them with my foot to stand them upright. Not sure how long this will be good for. The main bed of corn should also be hilled preventively, but I was too beat.

Also, I was dying to get the ragweed out of the asparagus forest next to the hives. I debated about how to dress for the job, but settled finally for just my bee veil and a very low profile, crawling cautiously on hands and knees along the bed with the pruners and clipping off the ragweed stems and leaf petioles. There turns out to be more than ragweed in that bed, mainly honeysuckle and violets. Some of the elephantine asparagus ferns were flopped over, making the job even more difficult. I couldn't see too well in the veil, either. Still, it looks better. Tomorrow I'll have to gather up the weed corpses; tonight enough was enough. Bees sizzled past my ears, but I kept low and was not menaced.

By the time I finished, the ducklings were peeping for their supper. I fed them, checked to be sure they had water, and essentially left them to their own devices for the night. A little while ago I went out, curious to see whether they'd put themselves to bed. They had. Ted had plugged in the light when he got home, and from what I could see through the cracks they were all inside, snuggled down in the increasingly filthy straw, asleep.

I wonder which predators we need to worry about. Could a raccoon get into that pen? With the roof laced to the sides, it's tight but not all that strong. We've seen raccoons around here, but have never lost any corn to them. I suppose ducklings might attract them, though. Apparently a raccoon can easily kill even an adult duck. But I don't think I can do any more about that than I've already done. Past a certain point, *que sera, sera.*

I can't *wait* to get back from Utah and into maintenance mode. Hard to believe the day will ever come. I wondered, sitting stupefied with fatigue on the patio before dinner, if this kind of overwhelmed-ness mightn't also be part of the homestead life—the reality that you can work incredibly hard and lose it all to weather, disease, grasshoppers. Raccoons can eat your ducks. Your fish can get cloudy eye; your Prima can get cedar apple rust and die. You can do *something* about these problems, if you try, but to a large extent they're beyond your ability to control, to determine the outcome of, especially if you voluntarily limit yourself to organic sorts of responses. Just being knowledgeable and working hard won't guarantee success. This year, for instance, the tomato harvest is unpromising and the cucumbers will never keep us in a year's worth of pickles from the look of them now. Mount Pinatubo's fault, not mine; but my failure if I can't figure out what to do in time.

The Romano beans have their first flowers. They'll be ready to pick, and we won't be here.

I've never gotten back to finish the blueberries. Tomorrow morning, come what may.

July 5. In the very nick of time, a really nice day. The weather, for one thing, was perfect: bright, cool, clear—a perfect summer day. For another, my activities added up to exactly the sort of homesteading day I'd hoped to have more often than not during this year, and have hardly managed to have at all—neither hectic labor nor distractions, but steady, pleasant, engaging work.

Since the ducks were quiet I decided to have breakfast before herding them, with Ted's help, into the garden. The idea is that if we set them to foraging while they're really hungry, they'll do it with a lot more zest. They heard us talking and started piping; don't know if they stayed under their house till we came out or were out already. But they went over the grass like greased lightning, bouncing off the electric fence wire without much complaint, and streamed into the garden where they started hunting *avidly* for slugs.

It cheered me up a lot to see that just since yesterday they'd obviously gotten the idea of SLUGS into their little brains. They poked and prodded between cabbages, in the mulch, under the leaves, finding lots and lots of slugs in places where it would have been very hard for us to look. When they got tired later on and wanted to come home, every one of them had a glopped-up bill. Slug control at last! Terrific little ducks!

My next chore was picking blueberries. Tedious as before, but more rewarding: in just over an hour I'd collected three pounds—more than twice the weight of the last pick in an equal amount of time. But those were much smaller berries; a lot of these were the really big ones, from the bushes I never got to the other day.

The store didn't have any potato mashers—nobody mashes potatoes by hand anymore, apparently—but the food processor did a great job of cutting the berries without shredding them, and the pectin package label was right: a full batch of jam works better than a halved batch. We had more than enough berries for a full one, which turned out first-rate and made eight recycled jelly jars' worth of jam. As Ted didn't want to hassle with canning, these jars will go into the freezer. There are enough blueberries left over that we should be able to make another batch Wednesday before we leave, and thus not miss out too badly on this harvest which, by coming two weeks late, will force us to go off despite our careful planning and leave part of the crop unripened and unpicked.

That pretty much took care of the morning. We spent the afternoon picking up the chaotic house in preparation for guests: my science-fiction friend Greg Frost and his significant other, Barbara Wilford, due at 5:30 for dinner, and Ted's daughter Alison, on a visit from Chicago, coming later to spend the night. Having guests expected was a great help, since nothing but external pressure was going to get us to tidy up the house, tired as we both always are. Now, having subdued the mess, I feel a lot better, as if control has to a large extent shifted back to me.

Just before hitting the shower I decided the straw litter under the duck house should be mucked out. There's a kid's garden rake in the garage, probably left by the previous owners and of no use whatever until now; but it turns out to be the perfect tool for that job. Raking the manure-y straw out and into the dirt cart was a snap, and so was hosing off the concrete: I just removed the cardboard from the near gable and stuck the hose through. The water roared right through the mesh floor. I let it dry for a while before stuffing the clean straw under. The dirty straw went to the new manure pile, and we built old newspaper litter into the compost to make things look a little tidier.

The ducks were a big hit; everybody enjoyed watching them run so desperately between garden and pen. Greg and Barbara loved the coleslaw. Altogether, a really good day.

July 6. I've yet to spot a fish in the pond, but I saw several tadpoles today. They seem to like to lurk in the skimpy vegetation right at the surface, and they're very large. Ted claims to have seen a fish, and the surface of the water bubbles and dimples as if they were feeding, but I haven't laid eyes on one and don't expect to till the water clears up.

I got up fairly late and didn't feel up to herding ducks before breakfast—a light rain was falling—so I took the easy route and fed them. Unfortunately I hadn't checked first with Ted, who'd been out to review the slug situation, which was fair-to-middling, and had steeled himself not to weaken and feed the ducks, but to wait till I got up to help take them to the garden. It hadn't been easy to hear them carrying on and do nothing about it; he was pretty cross with me.

When we did put them out there later they hunted slugs for a while (probably not so energetically as yesterday, when they were hungrier) before cuddling up together in the far corner of the potato patch on some straw, and falling asleep. There's a little tree growing in that corner, but I doubt it kept much rain off. But they don't care a bit.

After Greg and Barbara left last night, around 10:00, we shined a flashlight into the pen to see if the ducklings were under their house. They weren't. They were nestled up together on the concrete next to the garage, and were startled and upset by the light (something else done by the book). But when Ted checked again, still later, they'd moved inside. And they spent a fair amount of time inside today, in their house-cleaned quarters.

We took Alison to lunch in Media, after which she wanted to do some shopping. In one store that looked like they might sell them I inquired about candle molds. They didn't (candles yes, molds no), but another customer overhead me and said *she* had some molds that she'd been thinking of throwing away, and did I want them? So later on Ted and I drove over to her house in the truck and picked up a large box of candle-making paraphernalia: molds, paraffin in big heavy blocks, an unopened package of wicking, crystals of dye to color the paraffin with, other oddments I couldn't identify. We took it all. None of the molds is quite right—there's one of the Praying Hands!—but I'm happy to have them anyway.

Jan Purbrick and her daughter Claire were out in their yard when we got home, and came over to visit the ducklings. While we were standing around chatting, suddenly Jan said, "There are a lot of bees over there." I glanced toward the hives—and realized at once, with mounting horror, that there were indeed a lot of bees. Too many bees. Another building swarm.

I couldn't believe it. "It's too late! They're not supposed to swarm now! It's the sixth of July—the season's over!" I kept yelling things like that, but obviously the bees (like the ducks) hadn't read the same books I had. It was the second hive—the one made out of the divide and the first swarm, the one with all the honey. I hadn't looked in there for weeks? *months?* In any event, the air above was filled with bees in that characteristic formation, the front of the hive was aboil with them, whatever I thought about it they were swarming.

So I went in and got the binoculars, set a chair in the shade of the pin oak, and waited to see where the swarm would settle. But, oddly enough, by then the air seemed *less* full of bees than before, and then still less. The binoculars showed equal activity in front of both hives, what looked like orientation flights or robbing, but no longer any black mass on the front of the hive.

Presently even that had stopped. I got on my coverall and veil and went to have a closer look, in case the swarm had settled so close to the hive that it only *looked* as though they'd had second thoughts and gone back in. But there was no little aura of bees, there or anywhere else in the yard, to denote the presence of a swarm cluster, and I concluded that this must be a case where the queen hadn't come out and the rest went back in as well.

I also concluded that if they swarm now, they'll just have to swarm. Jim concurs. The honeyflow's over, he says; if a swarm takes off, there'll just be that much less of a drain on the supplies. He quoted an old saw: "A swarm in May / Is worth a cow and a bottle of hay; / A swarm in July / Is not worth a fly." Enough's enough. If they try it again tomorrow, bon voyage.

This evening Liam and both his parents came over for a lesson in how to take care of the ducklings while we're away. Liam's a serious and responsible kid, but he's only nine and he's never had any pets. We wanted to be sure he could work the switch on the hose nozzle, that he's tall enough to fill the waterer with it, and so on; and he needed to understand what the job entails. "We thought three dollars a day would be fair," I said. "One dollar per feeding. Is that okay with you?" Liam shrugged. "I guess. Sure." Lynn and Jim glanced at each other in a startled way and laughed; clearly they'd expected to hear less. "It's a lot of money, but it's a big responsibility," I told Liam. "These are little living creatures, and they'll be depending on you for everything." He nodded soberly, then grinned while I photographed him filling the waterer with the hose.

While the three Castellans were here, I made a mistake. I thought it would be fun for them to watch the ducks being herded, but the ducks were under their house and didn't want to come out. We managed to drive them out by poking the wands through the floor, etc., and they raced along obligingly enough. But it was dusk, so we had to turn them right around and herd them home again. A few more neighbors wandered by and joined the amused throng, and in all the commotion the ducklings got confused and scared. Instead of running, they wanted to hunker down. Too many people yelling and moving around—and none of this for their good, but for our amusement.

And I felt bad. I felt we were treating them without respect, as objects we could harry at our whim, and that this was absolutely not right. Ted agrees. We won't let anything like that happen again.

July 7. Yesterday the screenhouse contained three hills of one, one, and three melon sprouts respectively, and three empty hills. Today two of the group of three sprouts had keeled over with what looks like wilt, though I've never seen wilt at such an early stage before. These little plants have one true leaf smaller than their seed leaves and one tiny suggestion of another. I felt reluctant to invest much time or energy into making the screenhouse bug-tight, feeling that it may now be a case of shutting the barn door after the horse has escaped. But three plants is still three plants, so I did a halfhearted job of pulling the Reemay off the brussels sprout bed, cutting it to size, and sealing the ends with stones and boards on top, Reemay stuffed between stakes and screens on the sides, and dirt across the bottom, all the way around. This was an easy, quick fix, so if tomorrow all three survivors are wilted, too, nothing much has been lost—at least not beyond the original construction.

I've put *finish hilling potatoes* on at least three different lists of Things To Do, right up there with *close screenhouse.* Now that job is done too. I used soil from the pond excavation—topsoil saved in the metal trash can, plus some mixed soil from one of the piles in the jungle— and buried the bottoms of the stems in the second potato bed. Two more plants have bitten the dust this week, both with roots resting on air. What a *very* bad idea this was! But that's all I'm going to do to rectify the situation; if the harvest is poor, at least I know better than ever to mulch with straw again, and at least the surviving varieties appear to have excellent disease resistance.

The ducklings are beating up the potatoes somewhat in their passionate search for slugs, but a week of not being put into the garden should give the plants a chance to recover. They bite pieces out of the outer cabbage leaves, too, but at this point the cabbages can handle a lot of duckling bites.

In remorse about yesterday, today was Duckling Sensitivity Day. I moved them from pen to garden and back again only when *they* seemed to want to be moved. Tonight they got their supper in the garden and lay around contentedly by the gate for a while; but after *our* supper, while we were sitting at the table talking, a peremptory peeping was heard. The ducklings were standing by the gate, calling us.

We opened the garden gate and stood back while they dithered, snatched a quick, reassuring drink from the waterer, then walked hesitantly through the open gateway and started running like crazy in the general direction of their pen, we running behind with our wands of power. I yelled to Ted to let them go in by themselves if they would, I wanted to see if they knew where they were going by now. And sure enough, they got stuck briefly behind their gate, but figured out the error by themselves and peeled in. I hooked the gate shut while they drank some more reassuring water. By the time we'd finished putting away the carts and tools, they'd settled down contentedly on the concrete. "That *was* what they wanted," Ted said wonderingly. It was. They wanted to come home, and they'd said so.

I sighted my first fish today: a bluegill lipping something at the surface. Also a beautiful pale-blue dragonfly, that somehow found the pond, having come from God knows where. The Sagittaria appear to be hopelessly dead, also one of the cattails—growing but rotten at the waterline. Other plantings doing well.

It was a bright, cool, sunny day, perfect for harvesting onions. I lifted the whole crop this morning and left it to dry in the sun right on its own bed. A good-looking harvest, some very nice big bulbs and not too many maddeningly little ones. One of the smallest went into this evening's delicious three-lettuce chef's salad. The central leaves of all the lettuces are suspiciously small, especially the Black-Seeded Simpson; they'll have bolted and turned bitter by the time we get back—but the cabbages are harvestable *now*, so we simply switch salad greens. While I was working the ducklings poked around in the soft earth turned up when I pulled out the onions, clucking happily. Not many slugs left in the garden by now—only two gucky duckybills in the crowd.

July 8. This morning I spent a couple of minutes setting up a spillway for the pond, in the event of a lot of rain falling in a short time while we're away, and another hour picking two pounds of blueberries. I picked anything with any deep blue at all, thinking of the writers who tell you that underripe berries in suitable proportions make better jam because of their higher pectin content. Most of the rest should be able to hang on till we get back.

While berry picking, and reflecting upon the bad season the commercial growers are having (because of the late spring), I admitted to myself that all my fussing over having to worry about money during this year is kind of silly. Real farmers worry plenty about money. They live one year behind their expenses, instead of one year ahead: borrow money for seed and equipment, grow the crop and sell it, pay off their loans—then take out a new loan. Rather than using each year's crop money to buy next year's seed, they finish up even at the *end* of the season.

A homesteader grows things to live on, not to sell. But even a homesteader needs a grubstake, which I didn't have. My shortfall problem is like that of a farmer with no loan and insufficient capital, or that of a homesteader with no grubstake; but to mind having to finance the year from what I "grow" is dopey. So I'm up till 11:00 writing on this machine night after night; some farmer is up delivering calves or going over his books. Only the grubstaked homesteader sleeps secure.

Liam and Lynn came over so Liam could practice one more time. He understands about the hose and operates it well, and he seems very keen on the job, asking hesitantly if he was allowed to stay and talk to the ducks after he feeds them. I said sure, they like company. I referred to them as "ducklings" and he objected: "You should just say 'ducks' now, they're not ducklings any more." They're as big as seven little turkeys; I guess he's right. But when I said "Well, when they serve

an eight-week-old duck in a restaurant, they call that 'roast duckling,' "
a certain tension fell between Lynn and Liam and me. All the neighborhood kids are having a hard time with the idea that eventually we
plan to eat the ducks. I sympathize; but we are the predators in the lives
of these little birds, who were born to be preyed upon by something.

When we conveyed them to the garden this morning they swept
right through the gate, requiring almost no herding at all. And like yesterday, they poked around contentedly all morning, till the white oak's
shade had withdrawn and the tomato shade didn't amount to much
yet—around noon—and then let us know they wanted to go home. We
opened the gate and just stood back, and they walked through, then
started running.

The white oak with its stone surround is directly in the path
between the garden and the duck pen. They weaved desperately back
and forth as they approached this obstacle, then screeched to a halt, unable to decide which way to go around. Ted helped by giving them a
nudge, and they careened into their pen and made for the waterer.
They've learned! They know how to tell us what they want, and we
know how to respond! I really hate to leave now, fearing they'll double
in size and forget everything while we're gone.

I mucked out their house again, hosed off the concrete, and
stuffed in clean straw. They were in the pen when I was doing the hosing and were distressed by the sound; but when I stopped they wandered in to check things out. "What's that they're eating?" said Ted. We
couldn't tell. Seeds from the straw? Their own poop? All of a sudden I
realized it had to be sow bugs! I'd noticed lots of sow bugs under the
straw when I was pulling it out. Sow bugs into duck meat—almost as
good as slugs into duck meat.

The other chief task of the afternoon was collecting the onions
and spreading them out on screens set up in the garage, one across the
sawhorses and one across the back of the pickup. Too bad the garage
will have to be left closed up; otherwise it's an ideal onion-curing setup
following two days of bright sunshine. Rain was threatening, so right
after tea I gathered them all up, piled them into the cart—tails all point-

ing the same direction—and moved them under cover. The garage smells powerfully of onion, but that's one crop safely gathered in.

We tossed a few bread crumbs into the pond, but got no takers. I don't much like the look of the plantings. One floating heart—the one that had the two yellow flowers—looks poorly, or possibly dead. We lost the three Sagittaria and the cattail—green leaves withered today—and there's no sign at all of one of the Comanches. And the early burst of growth in the water clover and Odorata seems to be over. One heartening development is that the two surviving cattails are better anchored and no longer wobble in their pots.

I mind leaving the ducks, but not the pond. By the time we get back it should be easy to distinguish the living from the dead. Then I can call Lilypons with the death toll, they can send replacements, and we can get on with it.

July 18. We got back from Salt Lake City late last night, and—unable to wait till morning—went out into the sodden backyard with a flashlight to see if we still had ducks, and discover what had become of the pond and garden.

At the duck house there were peepings. (And quackings: when I called to tell Liam we were home, Lynn said, "Their voices are changing.") But mindful of the warning, in one of my books, that ducks are restive at night and agitated by lights, even moonlight, we didn't try to see under the duck house.

Instead it was on to the pond, where disappointment waited. Both Comanche water lilies appear to be dead; at least the leaf one had been sending up before we left had vanished. The floating heart is skimpy, and the plantings generally looked feeble. But the flashlight beam penetrated clear to the bottom, showing the underwater plantings, crates, containers, even the fish! I wonder now whether a flashlight beam wouldn't have done that all along, even when algae made things murkiest.

In the eight days since we left, the garden had *exploded.* There'd been a lot of rainstorms—an inch of water in the gauge—and the rain seems to have fueled a jungle growth of green. All the indeterminate tomato plants were enormous and had flopped sideways over the rims of their cages. *Lots* of little tomatoes on them, by the way. The sweet corn is very tall and tasseling; and the row I'd been fretting about, because I hadn't had time to brace it with soil, was standing as tall and straight as the hilled-up triangle patch. All the bean vines were covered with pods and flowers, and waving at the tops of their poles. The sweet potatoes were runnering at last. The big cabbage heads had split from all the rain. The potato bed I'd hilled just before leaving looked great—no more deaders, lots more foliage. And in the garage the onion leaves had wilted and dried.

All in all, what I saw was lots of work needed in every direction, but no disasters resulting from our week in the Rockies.

The trick (I told myself) will be to work steadily on one project at a time, refusing to get overwhelmed by the sheer volume of all the catch-up work to be done. I'm not going anywhere again till September; I've got six uninterrupted weeks before me. And now—exactly now— is when the payoff comes for the early arrival of the ducklings, which *don't* now have to be pampered and fussed over.

Things, however, did not get off to all that snappy a start. Every project was a struggle. After our cool arid week in the Utah mountains, it's mighty hot and muggy here in the Delaware Valley. And the effort could be partly mental, because something important has happened. There came a moment high on a mountain trail—spectacular vistas on every hand, cool thin air clear and sparkling, both of us grinning like fools—when Ted turned to me and said, with a gleam in his eye I've hardly ever seen there, "We could retire out *here!*" Not to Powder Mountain itself, but to Salt Lake City, where attractive little houses are so cheap by Philadelphia standards as to make our jaws drop, and seven canyons full of trail heads are minutes from Temple Square. There's a good-and-getting-better university handy, in case anybody feels like teaching a little, with a library whose holdings impressed us as extensive. (We looked ourselves up in the catalog and were very pleased with what we found.) We've even got a few old friends in town.

Once we'd recovered from the unexpectedness of this idea, it appealed to us both so much that we made a tentative decision to move out there in a couple of years. And the two visions—life in Utah, life on the homestead—may clash just a bit, till we get ourselves reoriented here. Of course, the whole scheme may seem unreal and/or crazy after we've been back awhile. I wouldn't be surprised. Ted dreads the prospect of being proselytized by Mormons night and day. (I tell him they don't do that, and they don't, but there's no denying it's their city.) On the other hand, can it be a coincidence that Utah is called the Beehive State?

Meanwhile, we have a homestead to run. I went out early—Ted was asleep—for my first look at the gigantic ducks. Though their heads are still covered with brown down, their wings, tails, backs, and breasts are partly to wholly feathered in black. And they *quack* now! Or some do; Ted tells me that only the females quack, that drakes have whispery voices. Maybe I'll vent-sex them soon; but maybe I won't, too. It doesn't yet matter much which sex they are.

A week of confinement in their pen has made a mucky mess of the place—grass all worn away, mudholes where the overflow from the waterer runs. I fed them; then I couldn't resist finding out whether they'd forgotten how to go to the garden. I opened the garden gate, then the gate to the pen, seized my wand, and stepped out of the way. And out they came, nervously, nibbling at the grass; and off they took, me racing behind, putting the wand to one side and then the other in an attempt to keep them more or less on course.

And, mirabile dictu, they remembered! At least, when they caught sight of the gate, they flowed through it just like old times—all but one, who got behind the V of the open gate and into the electric wire, and had to be caught and restored to the flock. I gave them some water and left them ecstatically scoffing down every green thing in sight (including the cabbages, which will have to be protected; what were cute little nibblings a week ago have turned into great big chomps).

The pond water is clear! You can see the bottom plainly now—oxygenating plants (which appear to be thriving), other containerized plants, snails, tadpoles, bluegills. The catfish are still keeping their low profile. The snails leave blunt, meandering tracks along the sides where they graze their way through the algae growing on the liner; the smaller, wigglier "tracks" must be from tadpole mouths.

Liz asked last night on the phone how the water hyacinths were doing. I was able to report that each had put out a daughter plant and produced a bristly bunch of roots. *They* are certainly the reason the water has cleared up already in spite of the requisite two-thirds of the surface not yet being covered by lily pads. We went out after lunch and threw pellets of bread to the bluegills, who snapped them up energetically till they'd had all they wanted. They're so little, it's hard to imagine any of them getting big enough to put on a plate, but if we feed them right . . . time to get the worm farm going.

I pulled some dead lily pads out of the water. They were definitely very dead. One of the Comanche tubers has some hard protuberances that might be leaf buds, but what I *think* is that they just didn't make it, from being so long out of water.

We looked about us for the most desperately needed harvesting activities and set to work. I picked blueberries: two pounds, two ounces, with probably no more than one pick to go before we can disassemble the netting system, and weed and mow the tangle of ornamental strawberry vines and grass that's grown up around the enclosure. Also, at Ted's request, I cut a cabbage—the one hardest hit by ducks—and he made coleslaw, finally getting the dressing almost right. We both picked Romano beans to marinate and team with lettuce from the health-food co-op—our lettuce, all three types, bolted while we were gone. And while Ted was making the salad, I went out to the garage with a pair of scissors and cut the leaves off the onions curing on the truck. We'll need it tomorrow as a truck, not a drying rack. We have to pick up supplies: lumber for the pond frame, fencing materials for the ducks' swimming pool and larger run.

The last thing I did was the duckmuck. The straw that came out from under their house was packed solid with manure—pretty smelly, but I'm surprised it wasn't worse, given the hot weather. The ducks stayed in the garden nearly all day. They gave every indication of having a great time, and the messy bills showed they'd worked for their keep a bit as well.

No gardening, not today.

July 19. There's a kind of skin on the surface of the pond—dust? Something dissolving off the liner? You can see it, wipe it with your hand, which then shows a reddish stain. The fish don't seem to mind. When Liz comes back, maybe she can tell me what it is.

The pickup, relieved of its burden of onions, went to the DIY store and brought home lumber, fencing, coarse sand, pea gravel, and some other odds and ends, including a replacement lure for the Japanese-beetle trap. The fourteen-foot two-by-twelves stuck way out over the end of the truck bed and had to be weighted with the fifty-pound bags of sand and pea gravel. We stapled Ted's T-shirt to the end of one board for a flag, and drove home very, very carefully.

The bathroom renovation had suggested an obvious solution to the problem of how to provide the ducks with water to swim in. Bill and Jeff finished the job while we were away, and carried the old bathtub out to the pen, where it's to become the ducks' swimming pool. Ted volunteered to dig a dry well for the tub to drain into. When the hole was big enough, he dumped in most of the bag of pea gravel; then we shoved the tub into position—it's cast iron and weighs a ton—and hosed it out. The water poured out the drain and disappeared into the pea gravel. A simple arrangement that works exactly as it's supposed to; now to see if the ducks will go swimming.

While Ted was digging, I baited the beetle trap and hung it on the clothesline, intending to use the catch for duck feed; but when I dumped the catch, a seething mass of beetles, out of the bag, the ducks eyed it warily and avoided it (and the beetles all flew away). The ducks had happily spent the whole day in the garden out of harm's way, catching slugs and munching cabbages; maybe they just weren't hungry. Try again tomorrow.

Dry well finished and functional, we set about enclosing the tub and covered pen inside a much larger uncovered run, using four-foot chicken wire with two-inch meshes. Ted made the happy discovery that an old wooden-framed window screen in the garage was exactly four feet tall and would serve perfectly for a gate, relieving him of the need to build one. One day this week I'll scrounge up some hinges and wire the wooden posts to the metal ones so the gate can be hung, and that part of the project will be finished.

The rest of it—rigging a ramp the ducks will climb to get into the tub—will take further thought. But except for that—hardly a minor technicality—we're all set.

Liam came over this morning to see the ducks and collect his duck-sitter's wages. Eventually I sent him off, because we had to leave to get the fencing. Later the Purbrick kids dropped by while we were working and I sent *them* home, saying they could come back tomorrow—but I think we're going to have to appoint official evening duck-visitation hours. Naturally the neighborhood kids feel a personal connection with the ducks and are interested; but I don't want to stagger forth at the crack of dawn, in the underpants and T-shirts I've been sleeping in, and find Ben Purbrick peering over the fence.

We've discovered that the bluegills will enthusiastically eat bread rolled into little pellets and dropped into the pond. The flakes that alight and float on the surface interest them less than the ones that sink below it. When we walk out to the pond now, an attentive little audience of bluegills is instantly before us, quiet in the water, facing us, waiting to be fed. They take about a quarter of a piece of bread's worth of pellets before beginning not to bother chasing after the tiny crumbs that dissolve off.

Two more dead snails; I sniffed them (definitely dead) and pitched them into the jungle. The irises and cattails are sending up healthy-looking new growth; no problem there. One floating heart produced another little one-day yellow flower, its third.

July 20. I got up late after a bad night and had to leave right away for town, so the ducks spent the whole hot, muggy day in their pen. As soon

as I got home, around 4:00, I fed them and herded them into the garden. I had a hard time getting them out but the wand eventually harried them through the double gates of the two pens, and they ran across
the grass, directly to and through the garden gate. Once inside they
chomped on some leaves, but it was obvious immediately that something wasn't to their liking. After I left, they stood at the gate and
quacked and quacked. I went out again to show them where their water
was, in case that was the problem (as it sometimes is), but no: quack
quack quack *quack*. Finally I gave up and let them out. They walked
straight through with no urging and ran right back to their pen, where
they played in the water and then went under the house, quiet at last.
They hadn't *wanted* to be in the garden, and they managed to let me
know that.

I was describing some such incident to my sister Jan on the
phone recently. Her response: "You *communicate* with your *dinner*?" She
went on to say she just could not possibly eat an animal she'd raised and
cared for, not that she was criticizing me.

I've got this issue all straight in my head, but we'll have to see
what happens when it's actually time to butcher four of these scalawags.
"They're so cute!" Ted said as we were putting them in the garden yesterday. Indeed. They do seem like one organism divided into seven
parts, an organism that can function only collectively. But collectively,
let's admit it, they're pretty cute.

I wrote *Pennterra* partly in order to see if I could devise, for another planet, a better system than natural selection as the mechanism
of evolution—improving on God, you might say. But here on Earth,
this system is the one we've got, and I understand that and accept it and
am trying to live by it. In four to six weeks we'll see how well I do.

When I went over to look at the pond this morning there was a
beautiful green dragonfly hunkered precariously on the skimpy water
clover foliage, its abdomen curved down into the water like a siphon.
Obviously laying eggs. While I stood watching her try to keep from
deep-sixing her inadequate platform, another dragonfly buzzed around.
The first flew up and tried again. This time she landed on the floating

heart, but that was no better. Then she tried the parrot's feather: still worse. Finally she lit on a lily pad—good footing—but in the center, too far for her ovipositor to reach into the water! She circled round and round, probing with her abdomen but always hitting lily pad, then tried another, smaller pad, and managed at last to achieve both firm footing and access to the water. So there will be dragonfly nymphs in the pond, unless the fish eat them all.

The rest of the onion tops are in the compost bin. The bare bulbs are drying on the screen set on sawhorses. In a day or two the whole harvest gets weighed and goes into the fridge. When I manage to get into the garden, I'll turn the empty onion bed again and plant the fall carrots that always do so much better than the spring ones.

Again this evening I tried to get the ducks to eat Japanese beetles. This time there was no doubt: I'd been feeding them bread, to lure them out from under their house, and put some of the bread in their feed pan, where I'd dumped the contents of the beetle bag. They'd been quacking and going for the bread like anything, but when the pieces hit the pan where the beetles were sorting themselves out and beginning to fly away, they became quiet and wouldn't go near it. One duck did eat one beetle that had crawled apart from the main mass, but this is clearly not a workable idea, unless I want to toss the beetles to them one at a time. (I don't.)

July 21. I called Lilypons this morning and read them the Death Toll: Comanche water lilies, both; Stellaria, all three; snails, six (and more keep floating up all the time); Cabomba, ten bunches. Also Myriophyllum, order five bunches short; and floating heart, puny and no new foliage, though still putting out the occasional flower.

The person I spoke to was very pleasant and cooperative, made a few helpful suggestions (feed the floating heart a Lilytab, set the Comanches six to eight inches below the surface, no deeper)—helpful, but annoying since the pond would have been pretty well established by now if they'd sent me a complete reorder weeks ago when I called to complain.

The Odorata is making new leaves steadily now, and the bog lilies and cattails (apart from the one that rotted, which, dammit, I forgot to mention to Lilypons) are growing well. Also, as far as I can tell, the underwater plants are all thriving. So it's not all bad news.

Then a cheering call came from Gordon Van Gelder, my editor at St. Martin's, with news about *Time, Like an Ever-Rolling Stream;* a royalty check of over three thousand dollars on its way, and a paperback sale to Del Rey—only five thousand, but it all adds up to a great improvement in the cash-flow situation.

July 22. The onions came in from the garage today, our first main-season crop. Most went into the crispers in the fridge, the overspill into a paper bag on the shelf. Total weight from two quarts of sets: 35 pounds. I thought that seemed pretty high and did a quick comparison with previous years, as follows:

1992	2 qt sets/Jung/$4.95	35 lb.
1990	2 qt sets/Jung/$?	25.5 lb.
1989	lb sets/Vermont Bean/$7.75	no weight recorded, but both crispers filled to the brim
1988	600 sets/no co. recorded/$?	filled 1½ crispers
1987	1 lb sets/Burpee/$5.50	between 12 and 13 lb.

Valid comparisons are hard to make with these shifting measurements of weight and volume—do 600 sets equal a quart? a pound? does a quart equal a pound?—but the volume of the harvest, measured in filled crispers, suggests that this is an unusually good year for onions. The 1990 and 1992 figures bear that out, especially because we thought we'd done well in 1990. Ted was convinced that his applications of Spray-N-Grow were responsible, but he never got around to spraying any Spray-N-Grow this year. The low harvest and small bulb size of '88 were certainly due to extraordinarily hot weather. This spring and early summer were as cool as '88 was hot—bad for tomatoes, good for onions. Things balance out.

We heard today that some birds did get under the blueberry netting while we were in Utah. One of our neighbors was extremely upset and angry: what business did we have setting up that bird trap and then going away and leaving it? Learning this made me sorry, but also thoughtful. When civilized people and wildlife occupy the same space, such things are bound to happen. However hard we try to make the setup bird-proof, for our sake and theirs, every year a few birds do get in. Last year we were home the whole time the netting was up, but while taking it down I found the skeleton and a few feathers of a catbird that had gotten itself entangled in the netting and died or been killed there without anybody noticing. Several of our neighbors have cats which kill birds quite regularly. The angry neighbor is a long-term patron of Chemlawn. Our scruffily maintained yard provides better habitat for birds than anybody's around, and we keep the feeder filled all winter. But it's still too bad.

Anyway, this year at least, all the trapped birds managed to find their own ways out.

July 24. Part of the reorder from Lilypons arrived last night—almost unbelievable speed. Today was cold and wet, but I went out anyway this afternoon, shucked my jeans, lowered myself into the water, and planted the contents of the carton: two Comanche water lilies, each with several leaves attached to nice, rubbery, obviously living stems, and three Sagittaria. I stepped on snails and nameless guck in the process, and several times Something (a tadpole? a catfish?) nibbled my toe, and lifting the heavy planters onto the shelf by myself wasn't easy. But all went well, except that by evening one Comanche had drifted free and will need to be replanted tomorrow.

I pulled out the dead Comanche tubers and tossed them into the bushes, from whence they sent up an unspeakably awful stench of rot. When Shakespeare said, "Lilies that fester smell far worse than weeds," he certainly hit the nail on the head.

Wading around in a gingerly way, I poked Lilytabs into all the containers and covered the soil of the new plantings with pea gravel.

Close to, the floating heart does seem to be sending up some new foliage. And I was delighted to discover that we have some teeny, tiny water striders, babies born to the ones that mysteriously turned up as soon as we'd filled the pond. I pitched bread pellets to the bluegills till their feeding frenzy slowed down, reflecting that I need to find out what else bluegills like to eat and provide that, at least till they get big enough for worms (and till I have some worms on hand to give them).

I'm starting to hate the ugly, unfinished look of those raw liner edges. Hope we can frame in the pond over the weekend.

The ducks spent almost the whole day in the garden, enjoying the wetness and the lack of sun and heat. Around teatime I heard quacking and came out to find them standing by the gate. When I opened it, out they came—and ran, without panic, across the yard, straight into their pen. All I had to do was close gates and serve supper. I *do* communicate with my dinner, or at least it communicates with me.

July 25. This morning when I went out to feed them, the ducks pecked at the feed dish in a perfunctory way but kept on quacking. On a hunch, I opened the gate. That was the ticket: they flowed over the grass and into the garden, whereupon the quacking ceased; no more was heard from the ducks till around 4:00, when I found them standing by the garden gate declaring that they wanted to go back to their pen. The same sequence in reverse. Very agreeable.

We got the "gate" (a.k.a. wood-framed screen) hung on the outer fence today. We couldn't find a hasp, but Ted made a perfectly satisfactory one out of wire and a cup hook. It was rainy, so we put off devising the ramp to the bathtub for another day. I feel some urgency about this, because pretty soon the ducks will be too big to fit into a bathtub all together, and they never do anything except all together. Their bodies are well feathered now; only their necks and flanks are still downy-brown. The glossy black feathers have a green glint, very snazzy. The wing coverts show as pinfeathers, like chrome trim at either side. Hard to believe they won't be "fully feathered" for three more weeks; they almost are now.

This afternoon we disassembled the blueberry-netting system and picked the last pint of berries. *Such* disagreeable stuff to work with, that netting; weeds had grown right through it, lacing it to the ground, and of course it snags on everything. I'm convinced now that the only way to deal with berries—blue, black, or rasp—is to cage them inside a permanent framework with removable screen panels, big enough to walk into for picking. There's a fine place for *caged* berries in the Ideal Homestead of the Future.

The bluegills like raw hamburger! I thought they might. We were having hamburgers for dinner and decided to take a little meat out to try on the bluegills, before frying up the rest. The fish went for each crumb the same headlong way they go for bread pellets. When they got filled up, Ted declared that slug-eating ducks would also eat hamburger and threw the rest in the duck pen; but the ducks got flustered and fled to the other end of the run, where they stayed as long as we had the patience to watch. We left them playing with some cabbage leaves I'd tossed in earlier and ignoring the meat.

The garden is in a dismal and desperate state, though I brought in, and we consumed, the first tomato: a bird-pecked Viva Italia, and very good it was. This evening we blanched and froze one pint of wax beans and four of Romano beans; but after a whole week back, I've yet to do any gardening beyond picking beans and decapitating cabbages. I'd meant to spend all of yesterday in the garden, but with the rain and chilly, cloudy day, the plants were too wet to work among without the risk of spreading disease. The huge, flopped-over tomatoes, which have grown steadily all this week as well, are pulling up the stakes that support their cages. One reason I took the blueberry system apart today was to recycle those stakes into the tomato beds as extra cage support. Let's hope tomorrow's weather doesn't stop me from getting that job done.

The whole garden is a riot of weeds—the usual July thing, made a good deal worse by two weeks of laissez-faire gardening. Tidying, weeding, and mowing are needed, in that order. Nut sedge everywhere, thistles all through the potatoes again . . . the electric-fence wire is touched by weeds in dozens of places and can't be packing much of a wallop, if any . . . the tomatoes should be pruned back, once they get better supported . . . etc., etc.

Still, everything looks fairly to extremely healthy, and very vigorous. (Ted says: thanks to the perpetual fertilizing of the ducks, and he may be right. They go everywhere, pooping as they go.)

A sign of how tired I am: when I went out at teatime to put the ducks back in their pen, I noticed that once again that stupid second hive was trying to swarm. I'd decided to let them go if they wanted to go, but did mean to go back and see whether or not they were actually leaving—and I *forgot* about it! I penned up the ducks and went in to tea, never giving the bees another thought.

How many times, though, has the hive tried that? How many times have some of the bees taken off? I can't believe I've managed to catch them in the act every single time—it's not like I'm out there looking all that often at this point in the season. I definitely need to requeen in the fall, if I can figure out how to do it.

There are fewer bees bearding the entrance this evening than in recent days, so maybe they made it this time—though if they did I saw no sign anywhere in this yard of the telltale whirling cloud. If it ever stops raining I'd better have a look in that hive, and I'd also better schedule a day for extracting, before the remaining forces eat all the honey. This winter we may be down to one hive again; I doubt the parent hive would get through on its own. I'll probably combine them.

Amazing how I hardly think about the bees these days. Everything's ducklings and fishpond, livestock that *won't* take care of itself if I don't tend to it. Maybe I've been given a sign: *Et in Arcadia apis.*

July 26. Today was the much-anticipated day when the ducks were finally introduced to their bathtub swimming-pool. For one of the seven it was a terrific success, for another a qualified success, for the rest an exercise in desperate frustration. For us there were both frustrations and hilarious delights.

We cobbled together a wide ramp out of more rough lumber

DUCK HOUSE LAYOUT

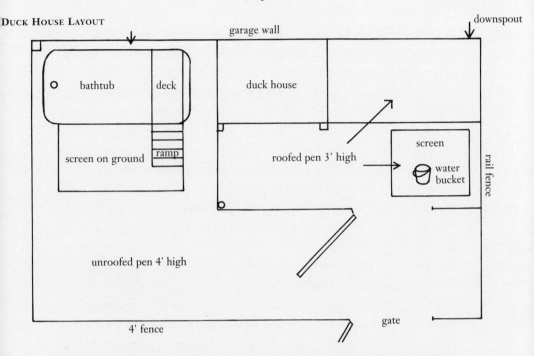

from the English shipping crate, and nailed cleats across to hold the pieces together. A board cut to fit across the back of the tub made a handy deck. But once everything was set up and the tub filled with the hose, the ducks simply weren't interested in sampling the pleasures of all our hard work on their behalf. What a letdown! I caught a random duck and put him/her in the tub; he/she scrambled frantically over the side and fell, whump! onto the ground—shocked but unhurt. I tried another with the same results. Then we resorted to locking them out of their inner pen, where the waterer is, and putting their feed on the cleats and the little waterer up on the deck, hoping thirst would drive them up.

For quite a while, no luck. A couple of the rascals would climb up a little way, then fall or jump off and stretch their necks like geese, trying to reach the feed high up on the ramp. Finally one particular duck got pretty good at walking up, and even down, the ramp. We'd put feed on the deck, and the duck would go up time after time and gobble it in lonely splendor while the others stood around quacking enviously. A second duck also learned to climb the ramp, but less often and with less confidence, descending by walking partway and then jumping the rest.

These two drank from the waterer, but also from the water in the tub. The overflow valve, unfortunately, kept the water level too far below the duckdeck for easy entry or I'm sure they'd have gone right in. The reason I'm sure is that we finally couldn't stand the piteous quacking for water of the five who wouldn't climb the ramp, and filled a dishpan for those five, to see what they'd do with a much smaller, more accessible "bathtub." What they did was surround the thing, drink thirstily for a bit, and then—one and all—submerge their entire heads, like so many miniature black ostriches! They'd never been able to do that before, but all understood that this was basic duck behavior. One little duck strained against the side of the pan, thrusting its head farther and farther underwater, scrambling at the bottom—and suddenly, *Geronimo!* it was over the edge and swimming! Almost at once another got over the edge as well, and between them they pretty well filled the dishpan. Great excitement for all concerned. They practiced jumping in and out—and then we had to get cracking on other projects, so I don't know what they did after that. But it was obvious that an in-ground pool would have posed no problems whatever—the elevation and ramp were the difficulties—and that a higher water level is needed in the tub.

From time to time we'd interrupt our work to give the ducks another lesson in ramp climbing. We dumped out the dishpan to keep their attention focused on the main object, and shut up the pen again, to make them thirsty enough to have some incentive besides food. Once, when Ted was tossing some bread scraps left over from fish feeding to the ducks, he got one to go in the water by tossing in a piece of bread. It clambered right back out again, but this was encouraging.

We worked with much concentration on the pond all afternoon, trying to get as much done as possible while the weekend lasted and Ted could help. So it wasn't till we'd called it a day and were loading the tools in the dirt cart that I glanced over and saw two black heads gliding back and forth in the tub. "They're in! They're swimming!" I yelled, and we hurried over (carefully, so as not to alarm anybody); but "they" turned out to be only two, almost certainly the two already seen to be more adroit than the others at ramp scaling. The remaining five were gathered around the tub, quacking disconsolately, stretching their necks to see in, at a total loss to imagine how their siblings had gotten into the water.

When a sudden movement alarmed them, the bigger duck got out of the tub easily, but the smaller had some trouble and needed several tries. Bill may be able to suggest a way to block off the overflow—silicone? putty? bubble gum?—to raise the water level. The problem of getting the five slow learners up that ramp remains, but we'll see what patience can accomplish.

We were delighted and chagrined, both, by the situation. The plight of the five frustrated ducks is comical but pathetic; this bathtub is as much their birthright as it's that of our Einstein of ducks and Einstein's smaller chum. What if some of them never figure out how to climb the ramp? What if they're already too old to learn?

Ducks need and deserve to swim. Swimmers versus nonswimmers might help us decide who gets to be dinner and who gets to breed.

The day's other major accomplishment was "leveling" (relatively) the pond rim and fitting the boards around it. I started by trimming the liner to six inches or so of overlap, then peeled the overlap back and draped it over boards laid across the pond, to get at the hardened clay underneath with hoe and rake. One long side of the pond had to be built up, its opposite side dug down. We tucked a few blueberry boards under the liner on the low side, for support in critical spots, then smoothed the liner back in place and laid the two-by-twelves on top of that. It's far from perfect, but I think it'll serve, like all Ted's carpentry jobs: rough and ready, but *solid* as the Rock of Gibraltar. Joining the corners together can be next weekend's chief project. Meantime, when I've got a jump on the gardening, I'll mow down all the pokeweeds; there's been a pokeweed population explosion out there. Then I'll rake the whole site smooth, add topsoil and compost, and throw some grass seed around.

Bill says the stuff on the pond's surface could be *jet exhaust.* Lots of that in Lester, where he lives, near the Philadelphia airport. A reddish-brown tinge, not oily. The recent almost-inch-and-a-half of rain drove it under, but now it's back, or more has replaced it. If this is coming out of the air, the whole garden is contaminated—and so much for organic fish-rearing, too. How to get the water analyzed without spending a lot of money? I'll ask at Rodale when we go up for GardenFest!

Before tackling the pond, I persuaded Ted to help me tackle the tomatoes. He drove stakes; I pruned and rearranged, lifted and hauled (and found a second red Viva Italia somewhere along the way). All the cages have now been straightened up and all the loosened stakes driven in deeper. But also the foliage on each of the indeterminate plants is a dense piled mat, highly resistant to air circulation, especially problematical in this protracted rainy spell. Once again I'm astounded at my capacity for forgetting how hopelessly inadequate these three-foot-high cages made of spindly two-inch chicken wire are to contain— let alone support—the gigantic tomato plants I want to grow inside them. Every year I deny, then capitulate to, the same facts.

On the other hand, every year we get enough tomatoes for eighty pints of spaghetti sauce, so the drawbacks to the system don't include an inadequate harvest, and that's probably why I go on like this in spite of what I know. These untidy, septoria-yellowed plants, for instance, after all my early kvetching about the cool weather, are *loaded* with big green tomatoes.

Quick garden catch-up: we have bell peppers on one plant, jalapeños on at least one other, and an in-ground eggplant started. I picked the second Jazzer cuke, a nice big one. Most of the cabbages have split, more than ever did before. It's the rain. Ted keeps making coleslaw, but we can't keep up with them; I suspect twenty was too many for us this year, given that Salarite isn't a "keeping" type. Luckily the ducks like cabbage.

The sweet corn is growing silks, but the popcorn's starting to tassel, too. It's important that they not cross-pollinate—or rather that the popcorn not pollinate the Butterfruit and combine with its shrunken-2 (supersweet) genes, or the supersweet corn won't be. That tall row of bean poles between the two plantings was put there on purpose as a buffer, but I've been relying more on the twenty-five-day difference between them and, as is often the case, they're coming to maturity much closer together than that.

If the rain would stop, I could get in and do something about the weeds.

July 27. This morning, when Ted let the ducks into their outer pen, one of them went straight up the ramp and into the tub! The tub water is much dirtier today than it was yesterday; it's certainly obvious why you can't keep ducks on a small ornamental pond.

The garden finally dried out enough for me to tackle the rampant jungle growth of weeds. I tore through like a bulldozer—cowdozer, anyway—pulling thistles and pokeweeds out of the soil heaped on the potato beds, nut sedge out of the paths and the cucumber bed, violets and ground ivy out of everywhere. There was so much nut sedge in one path that I got Ted to mow it, then added another six inches or so of leaves from the leaf mine (much depleted but not yet played out) over the whole thing. In a couple of hours I'd done a pretty good job of cleaning up the main garden, beds and paths both, and could hold things out of the way while Ted mowed the long, long grass in the two turf paths. Another hour on the back garden should have that looking as good as the rest. Surprising how little actual time it can take to do these jobs that loom so large and burdensome in prospect.

While tidying things up I decided to dig the volunteer potatoes, even though the plants hadn't died. The big volunteer in the pepper-and-eggplant bed I particularly wanted to get out of there, as the eggplant on whose space it most encroached was looking none too robust, much yellower than the rest. So I shoved a shovel straight down into the soil of the path—soft from the rains—and pushed down carefully on the handle—and up came treasure! Half a dozen nice-sized smooth-skinned potatoes plus several little ones. None of the other volunteers produced on the same scale, but by the time I'd dug them all we had some six pounds of zero-input potatoes, perfect except for two the millipedes had gotten into. We ate one millipede potato plus some of the littles for dinner, also some Romano beans.

I happened to glance over at the hives at one point and saw that the *parent* hive was displaying unusual activity: bees in great numbers crawling up the front above the entrance and lots of bees in the air, including some that were fairly high up. That hive has been so far under strength since its second swarm that I could hardly believe it was trying to swarm again; but either that was what was happening, or a new batch of bees was learning to fly, or robbing was under way. But, as before, this didn't last all that long; in fifteen minutes or so, things seemed like they'd returned to normal.

Then a little while later I looked over—and the *other* hive was doing the same thing! Exactly the same! Swarming? Robbing?? Flying lessons??? The activity didn't last any longer than it had in the parent hive. Either all my bees are swarming fools, or this is a type of activity commonly engaged in by bees during the summer; but the truth is, I haven't the foggiest notion what they were really up to.

Incidentally, there now appear to be about as many bees bearding the entrance of the second hive as before, cooling off on the porch of a summer evening. It seems that whatever they were doing this afternoon may be what they were doing last Saturday as well—not swarming, but something else.

On another subject entirely: in the *Harrowsmith Country Life Reader*, which I've been perusing, there's a piece about some of the people who tried full-scale homesteading years ago, and where they are now and what doing. A couple of the women say: a woman can't homestead alone. My own view is that most women could *keep going* alone, once the homestead was set up and functioning. I think that with maybe two years' grubstake, I could do pretty much anything short of building a house or a barn, or doing major repairs on either one. But it would be close to impossible for most women—me included—to set it all up

on their own in a year, or even in several years—especially if they had to support themselves completely at the same time.

This homestead is almost set up, almost functioning. But without Ted's muscles its fishpond would be tiny, its duck house much dinkier and probably charmless. Also, without Ted's willingness to take over the mortgage payments for the duration, I probably couldn't be doing this at all—not in Rose Valley, that's for sure.

And frankly, even with the said grubstake, and lots of time, I doubt that I'd *want* to do it if I were alone. (Though I suppose I might if I had children.)

July 28. Today, the first crisp, dry, sunny day since our return from Utah, was the one day this week I had to go in town. But I got in a couple of hours of work this morning before time to hike to the train.

The first priority was to suit up and cut the mock orange back from around the hives, then mow the grass between the hives and the apple trees, and finally fire up the smoker and look in to see how the bees were faring. I wasn't able to make a thorough inspection, but did find out what I most needed to know: that the top super on the swarm hive is fully capped now. It's time to extract. I've named Sunday, August 2, as Extracting Day and invited Liz over to help if she likes (it's *her* ex-extractor!)—meaning I'd better get it up from the basement and cleaned and be sure I've got all the other necessities on hand: Bee Go, fume board, bee escape, honey containers, cheesecloth for straining, screens, whatever. Back to the books, in my spare moments, if any.

I also started clearing the electric fence of weeds. What this amounts to is disconnecting the fence from the battery and solar panel, then crawling on my hands and knees around the entire perimeter of the garden, pulling up thistles and violets, cutting grass with the clippers, straightening the plastic mulch and tucking it under the rabbit-proof chicken-wire strip. I cleared the whole north side beneath the bee barrier and started on the east side before having to quit.

Ted says two of the ducks spent much of the afternoon swim-

ming in their pool, while the others stood around the tub craning their necks and wondering how the Sam Hill those lucky ducks got up there.

July 29. Eight more bunches of Myriophyllum arrived from Lilypons and have been planted in a pot and sunk in the pond. Liz thinks the reddish brown skin is stuff from the dead lily pads, a very tenable hypothesis and a much more agreeable thought than jet exhaust.

I was throwing crumbs of raw hamburger to the fish at lunchtime when I saw a catfish—the first one I've laid eyes on since the day I put them in the pond. It was skittering across the bottom—hunting for lunch, I suppose—and made no attempt to compete with the bluegills for the hamburger I was flicking in to them. Nice to see it; I wonder where the three others are. There's actually enough dirt on the bottom, washed over from planting and fallen in during my digging last weekend, that more cats might have rooted contentedly around in that without disturbing the dirt in the planters . . . but it doesn't matter, and the bluegills are a lot more entertaining.

The Comanche I'd put on an overturned crate, following the suggestion of the Lilypons person, was trailing leaves on very long stems across the pond surface. Obviously it used to be deeper, so while I was planting Myriophylla I set it on the bottom.

Weeding under the electric-fence wire continues. This morning I finally cut the extra fiberglass post in half, fitted the halves with clips, and drove them in where they would do the most good. I'd worked my way beyond the screenhouse by the time the sun caught up with me. After tea I made it to the west side of the garden, but finishing there requires that Ted go through first with the clippers and beat the jungle of rose of Sharon and assorted wild shrubs into submission.

The ducks don't seem to want to go in the garden anymore. In hot weather they like their pen, they like to go under their house and stay there. One difficulty is that they keep getting separated by the inner fence and quack desperately till I go out and herd one group around to join the other. I removed the screen so they could go under the house

directly from the tub area, but no: they settled down instead under the ramp!

We can keep them at work in the garden by not feeding them in the morning, make them rustle up their own grub for a while first; but I didn't do that today. When the garden was all in shade, late this afternoon, I did put them in for a time while I was clearing the electric-fence line. Once in, they set about foraging with what appeared to be a will. I was singing to myself as I worked my way along—cowboy songs, for some reason—and after a while I noticed that the ducks had all come over to the other side of the fence and settled down. When I moved away, still singing, they got up and followed, and settled down again near where I was working. Very flattering; who else would follow me around to hear my rendition of "Ghost Riders in the Sky?" When I finished and went to dump the weeds in the compost bin, they clamored immediately to be put back in their pen.

Mexican bean beetles have hit for the first time since 1988. One pole of wax beans only so far. I was picking those and noticed the lacy leaves and the yellow larvae—pretty good-sized already. The *O.G. Encyclopedia* doesn't recommend Bt for Mexican bean beetles, so I hand-picked all the larvae I could find and took them to the duck pen. Two more pints of wax beans in the freezer.

With my last few ounces of strength I pulled up the tall, duck-stripped stalks of bolted lettuce and turned that half of the first back-garden bed. Ted had mined me some more compost from under the brush piles he's been burning—the old supply was completely exhausted—but I wasn't able to do more than turn the bed once before pooping out. Tomorrow.

July 30. Wrong call: today, instead of finishing the digging and weeding, I finished clearing the electric fence. Ted had planned to go in town today, but his meeting was called off so he stayed home, hacked the shrub jungle back a couple of feet, and mowed the grass. I followed and got that tedious job completed. The whole perimeter of the garden is now enclosed by a wire not touched by one single blade or leaf. The smoothed-out strip of plastic mulch underneath is anchored on one side by the chicken wire at the bottom of the fencing and on the other by firewood logs, grass clippings, blueberry boards—whatever worked. With a little vigilance it should be possible to stay on top of the situation now. Major maintenance shouldn't be required again this year, if we don't go away any more.

I looked for more bean beetle larvae in vain; but this afternoon I spotted the first striped cucumber beetle of the season, on the smaller Jazzer plant. Great. I hand-picked this one and squashed it, but if they're going to be around now I may decide to spray with rotenone. I don't like to—it kills beneficial insects as effectively as pests—but with only two cuke plants left on the trellis, I may do it anyway. On the insect front, I can report further that I saw a lacewing on something today, corn? beans?—good news—and that the other day I plucked a tomato hornworm off a potato plant and flipped it into the duck pen, where one of the ducks instantly pounced on it.

Three apples on the ground under the Liberty tree today. I brought them in. So few fruits of any kind, let alone undamaged ones, are left on any of the trees that I think we must regard apples generally as one of our failures, and not because I didn't spray. Next winter I won't prune a blessed one of those trees. There's not much else to try! Despite Ted's not putting any grass clippings on them this year they've all grown like weeds anyway. The Prima will probably die of rust. I'm demoralized on the subject of apples, that being one crop I've put a lot of effort into with poor results I don't understand the reasons for.

Pulling the crate out from under the Comanche lily stirred up a lot of bottom guck that has been floating on the surface all day. If it doesn't rain tomorrow I think I'll spray the surface with the hose for a while. A fine spray would dissipate the chlorine and aerate the water, and top up the pond while we're about it—no rain all week has lowered the level. The fish don't appear to mind, and it doesn't smell bad, but I want to clear it up if only for aesthetic reasons.

I walked around with the bag of Vermont Organic Fertilizer

and fed everything that seemed to need feeding. This excluded the tomatoes, beans, and cabbages, and the Butterfruit corn under plastic (which is too far along to profit from it anyway), but included pretty much everything else. Several more tomatoes are starting to ripen. Eggplants and peppers are coming on, and I picked a couple more cucumbers. Before we know it the main harvest will be rolling in. Got to finish catching up before it does! But even more urgently, got to take some time off from the relentless pressure of this work.

July 31. Which will explain why, when I awoke this morning to peals of thunder, I felt so happy. A morning off! The storm, which delivered two inches of rain in one single morning, fixed it so I could work neither on the computer nor outside, and I enjoyed a guilt-free interval of reading Sue Hubbell's bee book and making a few long-deferred social—as opposed to business-related—phone calls. This afternoon the rain stopped, leaving behind a saturated world of tropical heat and humidity. It was hard cranking up to do anything, and I settled for just siphoning the water out of the duck tub—mud colored and totally opaque—and refilling it with (briefly) clean water. "Just when we had it like we liked it!" they quacked, when they came back from the garden and saw what I'd done.

This first siphoning was a cinch. Following instructions in my Logsdon pond book, I unscrewed the nozzle from the garden hose, turned it on, and stuck it into the tub, along with one end of the empty hose I was using as a siphon. Holding the two ends underwater, I butted the flowing stream from the garden hose up against the end of the siphon hose. When water started running out of the other end, I removed the garden hose and dropped the end of the siphon hose into the bottom of the tub next to the drain. The tub sits higher than the lawn and the procedure worked perfectly. (I must ordinarily try to do this when the grass *needs* watering.) After the tub was almost drained I used the garden hose to flush the last half-inch of water and bottom layer of mud into the dry well, put in the brand-new tub stopper we bought yesterday, and hung the hose over the rim until the tub had filled.

August

August 1. Ellen Harris, my editor at Del Rey, brought her boyfriend, Eric, and herself down from Brooklyn today to see the homestead. It was a brilliant, crisp, dry day, cool and windy: wonderful! Wonderful also, as a change from battling problems and struggling to do too much, just to stand aside and let somebody else admire what we've accomplished here.

After lunch at the picnic table, featuring sandwiches from Boston Chicken and homemade coleslaw and marinated veggies from the garden, I suited up and opened the second hive to put the bee escape—a one-way gate that lets bees out but not back in—under the top two supers, something that would have been done yesterday except for the rain. Ellen and Eric watched with binoculars from the shade of the pin oak, just like Rika did. Everything went fine for a while, the frames were fully capped and ready for harvesting, and the bees were very sweet-tempered when I first opened the hive. But once the bee escape had been fitted into the hole in the inner cover, I had to position the cover under the supers I wanted to clear of bees. To do that, I had to take the supers off. And each of them was so heavy that, setting the first one back on the hive, I thunked it down pretty hard and made the bees mad. Unfortunately my smoker also chose that moment to burn out.

I hadn't bothered to change my shoes for boots, thinking I wouldn't have the hive open long. But when I jostled the hive an angry cloud swarmed up, and a bee flew up the leg of my coverall and stung me in the calf. I knew better than to slap at it, but did anyway, reflexively. Then I had to stop what I was doing, go in, scrape out the stinger, and put the boots on after all. Too bad to have an audience for this misadventure, the only blot on an otherwise enjoyable and gratifying day.

BEE ESCAPE

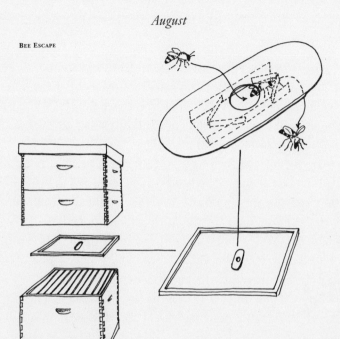

Eric seemed very, very interested in everything and asked lots and lots of questions—just the sort of visitor we hope to see more of.

August 2. One of the Homestead Year's Big Days: EXTRACTING DAY!

I spent the whole morning getting ready: washing out the extractor and assembling jars, lids, cheesecloth for filtering, the canner to use as an uncapping tub. I found and read the instructions for wielding the electric uncapping knife, ditto the extractor instructions. The fume board and bottle of Bee Go were located and made ready to use. I put the two spare empty supers in the garden cart, in case I had to move the frames into them one by one (if the bee escape hadn't worked), and a damp towel. When Liz arrived at 1:00 I was still engaged in organizing all this stuff.

While rushing about this morning, I'd noticed some unusual activity at the hives and gone over to investigate. The swarm hive had chosen this very morning to start robbing the parent hive. A gang of bees was trying to get in under the cocked-up cover in back, and a lot more were attacking at the entrance. I could see fighting. Great. In a way it almost didn't matter. Since I'm going to have to unite these colonies in the fall, any honey the second hive robs from the first will be given back to the first in the end. Still, it's a bad practice with high bee mortality, to be stopped if possible.

I remembered reading about a beekeeper who set a sprinkler on top of the hive being robbed and turned it on; this was said to be the only thing—other than a rainstorm—that would stop robbing once it had started. I haven't got a sprinkler, but I have got a hose; I dragged it over there, set the nozzle on a fine spray, stuck the nozzle through the broken brick, and took aim. Before letting fly I shoved an entrance reducer into the entrance so the guard bees would have a much smaller area to defend. The inner cover on that hive is screened; by turning it over, and thereby aiming the half-moon port upward, I eliminated rear access altogether. I also lowered the cocked outer cover. Those moves, made to discourage the group of robbers trying to get in the back way, are made at the potential cost of creating a ventilation problem for the hive. But it's a short-term strategy; for a day or two they'll be okay, and it did stop the robbing.

Shortly after Liz's arrival I suited up again and went out, dragging my cart behind me, to take the two supers off the swarm hive. Sometimes a bee escape clears a super in twenty-four hours, sometimes not. This time it hadn't worked, not well enough. After twenty-four hours, there were still plenty of bees left to buzz at me when I took off the telescoping cover.

That meant falling back on chemistry. I would have to drive the bees down out of the supers by using the fume board and the Bee Go. I was prepared; I had the board and my bottle of Bee Go with me. The bottle had a pointed cap, like the one on a tin of 3-In-One Oil, taped to it with strapping tape. The childproof cap was also shrouded in tape. Sweating in my coverall, I pulled off one gauntlet and managed to rip off the tape; but the childproof cap was almost proof against me as well, and in my struggles I managed to squirt some Bee Go onto my bare right hand.

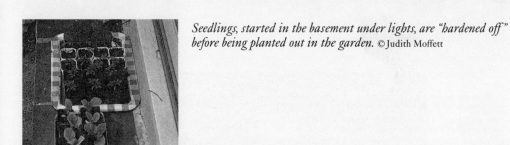

Seedlings, started in the basement under lights, are "hardened off" before being planted out in the garden. © Judith Moffett

The summer garden.
© Judith Moffett

Onion harvest: the first crop comes in. © Judith Moffett

The old newspaper trick.
© Judith Moffett

Working a hive.
© Liz Ball

A frame of honeycomb in spring: capped in honey in upper right corner, capped brood in center, worker bees everywhere.
© Liz Ball

The hot uncapping knife sliced through the wax cappings.
© Liz Ball

Extracted honey pours out of the honey gate.
© Liz Ball

Fuzzy-headed ducklings, pinfeath-ers just growing in, forage in the garden. © Judith Moffett

Swimming practice.
© Judith Moffett

Ducks can be "herded" with a wand.
© Liz Ball

*The Seven Samurai hustle
toward the garden's open gate.*
© Judith Moffett

*The author and six tempermental models contemplate
each other. In the background: bee barrier and hive.* © Ted Irving

*June 1993: Frank mopes while Winnie broods, her bill just
visible under the duckhouse, atop her nest of straw.* © Judith Moffett

The pick-and-shovel approach to pond excavation. © Judith Moffett

The author neatens the pond liner as water pours in. © Judith Moffett

Ted tosses shrimp pellets and the water churns. © Judith Moffett

*Spring 1993: bluegills,
having survived a month
under ice, come to be fed.*
© Judith Moffett

August 1993: the catch.
© Judith Moffett

A bucket of frozen Norway maple sap hangs on its spile.
© Judith Moffett

The author skims niter off the surface of the boiling sap.
© Judith Moffett

Hardy kiwis, bloom-ing on the pergola, make dappled shade and small, green, grape-sized fruits.
© Judith Moffett

This low-tech "blue-berry system" uses gar-den stakes, wire arches, and bird netting.
© Judith Moffett

The stuff is incredibly volatile and foul beyond description—that's why it works, of course. Instructions on the bottle read that if you get it on your skin you should flush the affected area with water immediately for at least fifteen minutes. Adrenalized as I was, there was no way I was going to do that, even if the stuff does "irritate skin." I held my hand under the faucet for maybe half a minute, then dried it off, put my glove back on, and went back to work.

I squirted a little Bee Go onto the fume board and put the board on top of the hive. You're supposed to wait two to three minutes, and I did. A roar was coming from the bees driven down from the top super into the one below. When I took off the fume board and glanced between the frames, it looked pretty empty in there. I pulled the first frame and found only a few bees, plunged halfway into cells of honey. I brushed them off, set that frame into one of the empty supers in the cart, covered it with a damp towel, and pulled the next frame. I did this nine times, till the top super was empty and all its frames, nearly free of bees, were in the cart under the towel.

I put the fume board back atop the second super on the hive, and while it was driving the bees down I dragged the cart back to the house. Liz had spread some newspapers on the floor by the patio door. She slid the screen back; I pulled aside the wet towel, heaved up the super, and plunked it down on the newspapers.

Then I had to restart the fire in my smoker, which had gone out again; after getting stung yesterday, I didn't want to be without a working smoker. So altogether I'd estimate that the fume board was on the second super more like five minutes than two or three.

When I got back with my huffing smoker and the cart, the roar from the hive was terrific. Only then did I realize, with a flush of horrified guilt, that I'd forgotten to take the bee escape out of the hole in the inner cover. Most of the bees had been trapped in the super with the Bee Go fumes like an audience in a burning theater, unable to get out through the one tiny exit.

I snatched off the fume board; bees boiled out. I lifted the whole super off and set it on the telescoping cover, which was upside down in the grass. "I'm sorry, I'm sorry!" I moaned to the bees. As if stealing their honey weren't bad enough! That unintended torture, the product of my forgetfulness, was no way to recompense them, and I felt truly terrible. Perhaps in consequence, and for the first time, I also suddenly felt a wrench at robbing them of their honey. Never mind that they'd spent the morning trying to rob their siblings; that's just being bees. What I was doing was unnatural and not in the bees' best interests. I thought about the Terramycin and Fumadil and menthol I'd dosed them with to keep them in good health, but I still felt bad.

But I proceeded anyway to brush the few remaining bees off each frame and tuck the cleared frames into the empty super in the cart, under the towel. When all were secured I popped the bee escape out of the inner cover and put the hive back together, hoping the house bees I'd brushed off, that had never yet left the hive in their lives, would smell their way home.

Back at the house, Liz had been keeping busy releasing the few bees that had come in with the honey. Some she herded out with a folded newspaper; others she captured with a glass and card and put out. I brought in a whole new batch with the second super, but she dealt with these while I was getting out of my veil and coverall, and fairly soon we were bee-free.

Indoors, that is. Out on the patio an outraged horde had gathered and was assaulting the screen. Reluctantly we closed the door to the patio, shutting off the smell of stolen honey. But as soon as we had started uncapping and extracting, more bees gathered at the kitchen windows, casements that couldn't be closed without opening the inside screens. So all afternoon the relentless, accusatory buzzing of bees accompanied our labors.

Ted had let it be known that this was one project he'd just as soon not be involved in and had busied himself with other matters all morning. He'd wound up being drawn much further than expected into the duck and fishpond enterprises, but he'd gotten interested in them, and he's never shown much interest in the hives. Yet when he came out for a few minutes to watch us work, he found extracting honey—the process itself—to be much more fascinating than he'd ever found the bees to be. So all afternoon I had not one helper but two, and that made a big difference in how long it took to do the job.

At Liz's suggestion—she'd done this before—we covered the kitchen floor with newspapers. I'd tied some cheesecloth across the top of the "uncapping tub" (we're using the canner, the biggest pot we've got), lashed a board, through which I'd driven a nail, across the top of that, and set the tub on a chair. When we were ready to start, I plugged in the uncapping knife and stood a frame of honey on the point of the nail. Cutting off the wax cappings turned out to be exactly like the many descriptions I've read of the process: you just saw the hot knife back and forth through the comb, using the top and bottom bars of the frame as a guide, then turn and repeat for the other side. The wax peels off in a sheet and falls into the cheesecloth, and there's the exposed comb, bleeding honey.

You stick the uncapped frame on end into the basket of the extractor and uncap a second frame, poking with the knife at low places that don't cut off cleanly. When there are two frames in the extractor, you close the lids and turn the handle. The basket spins, throwing honey against the sides of the tank. You crank till half the honey, more or less, is out of the first side, then stop and reverse the frames, and spin all the honey out of the second side. Then you reverse again and finish emptying the first side. The extracted honey slides down and collects in the bottom of the tank. You take out the emptied frames, put them back in the super, and keep going till all frames have been extracted. Simple.

It *is* fairly simple. I sustained a couple of work-related injuries: cuts on my fingers from washing the extractor and a gouge taken out of another finger when I tried to grab the free-spinning crank (dumb mistake). But in a couple of hours we had extracted all the honey from both supers and were having tea and English muffins topped with—guess what—lots of beautiful, light-hued raw honey. It's very light, almost lemon-colored, with a light flavor in contrast with the honey bees are said to make at the end of the season. Rather like maple syrup, which is also (ordinarily) lighter colored early in the season and more "robust" as spring comes on. I don't know how much honey there is in the tank and see no way of weighing it before we bottle it tomorrow, so the yield will be an estimate, but there are a couple of gallons of the stuff at the very least. A superb homestead harvest, a year's supply plus lots left over to give to friends.

What *isn't* simple is trying to separate the beeswax from the honey and save it. Cappings are the prime source of wax for candles, but what's left in the cheesecloth after the honey drains out of its own accord is still full of honey. We put the cappings into a double boiler and heated them till the wax liquefied and floated above the honey— discovering in this way that cut-off cappings are mostly honey instead of mostly wax. When the wax had cooled and solidified we had about half a coffee can of yellow beeswax for candle making, not much considering that procuring it was by far the most bothersome part of the whole process. Extracting would be a snap if we decided we could live with the thought of not saving the wax. It could come to that; but first we need to make a candle or six and find out how valuable beeswax really might be to us.

The kitchen and dining room are covered with stickiness. I keep forgetting about my injuries and washing honey off my hands, which hurts like the dickens. Tomorrow morning, first thing, I'll take the tubs and buckets and the spinner basket out to the hives for the bees to clean up. The honey will sit in the tank overnight, so the flakes of wax and other impurities can float to the surface. Tomorrow we bottle. Tonight we watch the Olympics and ignore the mess. My hand, by the way, appears undamaged by the Bee Go encounter.

This year, the second super started the season with ten frames of undrawn foundation. Next year only nine will go back onto the hive, in a super with a nine-frame spacer, and the bees will draw the comb out a bit farther on each face, making it easier to uncap.

August 3. Overnight the wax in the extractor rose through the honey and was floating on the surface when we got up this morning and peeked under the lid. After breakfast we assembled all the empty canning jars in the house, plus assorted glass food jars saved over the past year against precisely this necessity. Substances strong in essence (peanut butter, pickles, spaghetti sauce) contaminate the gaskets of their lids beyond even a dishwasher's power to remove the taint, so the collection is limited to jars that once held milder substances, like mayonnaise and, yes, honey.

Ted fetched and carried, I sat on the floor and lifted and lowered the honey gate, releasing a thick lemon-colored stream of honey into the jars and cutting it off again when they were full. A certain amount of dripping and spillage seemed inevitable—it's a good thing honey is soluble in water. Drawn off from the bottom as it was, the stream stayed almost clear for quite a while. Then suddenly clots of wax began to appear, and that was our signal to stop and filter.

Once we started filtering, very little happened for a long, long time; I was reminded in fact of filtering the maple syrup, another slow-motion homesteading procedure. My filter in this instance was a colander lined with four thicknesses of cheesecloth and set in a plastic bucket right beneath the honey gate. Honey fell in a thick rope from the extractor gate into the colander, filling it nearly to the brim. It then seeped through the cheesecloth and the holes in the colander, and fell drop by drop into the pail—but very, very slowly. We watched for a while and the level hardly sank. Plainly the thing to do was go on about our business, returning now and then to top off the filter, which did in fact become less full over time when not under observation.

So that's what we did. It created an awkwardness; Ted's son Terry and granddaughter Megan were coming, and we'd wanted to clean the extracting mess away before they arrived. But that wasn't possible, so the sticky newspapers stayed on the floor and the extractor stayed in the middle of the kitchen. We went out for lunch and brought take-out home for dinner. And all day long the extractor stood in the kitchen while honey slowly dripped out of the colander, into the pail.

At bedtime I rather messily filled two jars from this pail, using its molded spout (not a precision tool) to lower the level so the colander's bottom wouldn't be sitting in filtered honey by morning. By then there wasn't much left in the tank but beeswax.

The bees were still so riled up this morning that when Terry carried his honey-doused pancakes out to the picnic table, they chased him back inside. They'd discovered him right away and wanted to argue about whose honey it was. After breakfast I took the buckets and the spinner basket out and left them by the hives, and when I went back at dusk the bees had completely cleaned them up: every surface perfectly dry and clean, nothing left of the sticky mess that had been inside but some dry flakes of wax.

Megan has had a *fantastic* day, one I bet she remembers all her life (she's not quite eight), feeding ducks and fish and eating honey. She's not crazy about the *idea* of the bees, but she loves their honey—also the blueberry jam she put on her toast at teatime. This honey is so delicately flavored you can eat it with a spoon, and we all did. I've no idea how much of it I managed to consume during the past two days without getting cloyed out, but plenty.

August 4. Today the almost-empty extractor tank went out to the bees and came back as clean and dry as the other stuff did. All morning there was a buzzing cloud around the tank, but I only had to shoo off one bee when I went to bring it in.

Megan had another great day helping Ted, picking beans in the garden (and eating them raw), throwing bread to the ducks and fish flakes to the fish. This evening while I bottled the last of the honey, which had all drained through the filter at last, she sat on the kitchen floor holding the spatula and watching. When the pail was as empty as it was going to be, she held it in her lap and scraped the residue out with the gucked-up spatula till even she had finally had enough—this after two days of unrestricted consumption. There was honey on her legs, hands, arms—everywhere. I suggested a bath and got no objections at all.

The newspapers are finally up out of the floor, but everything's still sticky. The house gets cleaned on Thursday—not a moment too soon.

August 5. My main project for this day was to wash all the honey-extracting paraphernalia and get it put away. My secondary project was to wash the honey off the outside of all the jars and get *them* put away. For some reason honey is weighed in pounds, not fluid ounces. We tried to figure up our approximate total poundage from the two supers by getting a net weight for a one-pint jar (21 ounces—honey is heavier than water) and estimating how many pints we put away. It comes to fifty pounds—more or less—of light, clear, pure, perfectly beautiful homestead product. About twice what we use for household purposes in a year, so there's lots of surplus for gifts.

I've gotten over feeling bad about robbing the bees. Now I just feel tremendous satisfaction.

I drove the reluctant ducks to the garden this afternoon, lest they become fat and slothful for lack of foraging exercise and cabbage vitamins. Once there they were perfectly happy, rooting and clucking while I dug a full garbage can of compost into the former lettuce and onion beds and raked them smooth. The ducks don't like being stared at or approached, but as long as I don't pay them any direct attention they like having me around—something not mentioned in any of my duck literature.

The Mexican bean beetles are worse. I hand-picked some more larvae, like little yellow pincushions, but the situation has gotten beyond the power of hand picking to control. I'm going to have to spray. The loss of the first Bush Baby cucumber plant, just as it was beginning to grow a few little cukes, is another good reason if I needed one to break out the liquid rotenone; but I hate to use those broad-spectrum insecticides, however organic they are, and in past years have lost melons and cucumbers by the score in consequence of this reluctance.

This year, let's fight back. Rob Cardillo, a photographer from

Organic Gardening magazine, is coming tomorrow to photograph the ducks for a piece they're doing on poultry. I don't want either to limit the garden space available for his purposes or to expose the ducks to rotenone sprayed less than twenty-four hours before, even if it's supposed to be okay. But tomorrow evening, when the bees have gone to bed, out comes the sprayer.

This afternoon I picked, blanched, and froze two pints of wax beans, also two of regular green Kentucky Wonders, which somehow always get mixed into the packets of Kentucky Wonder wax bean seed (or are they genetic throwbacks?). Both batches were somewhat over the hill. Hard to keep up with beans.

We're getting a little tired of coleslaw; good thing the ducks are so nuts about cabbage. Every time we take them to the garden they flow through the gate and head straight for the cabbage bed. Last week, at my request, my friend Vicki—a Joycean—supplied me with a recipe from the *James Joyce Cookbook* for a casserole using the ingredients we've already got on hand: potatoes, cabbage, onions. This savoy-type cabbage is pretty bland when cooked and the casserole resembled mashed potatoes more than anything else, but the next day we added chicken stock to the leftovers, which turned them into a pretty good soup. Next time we'd better use at least twice as much cabbage if we want to approximate the intention of the Irish original.

The first ripe Gurney Girl, in tonight's salad (with marinated Jazzer cucumbers), was disappointing. Early-season blandness? Don't know; but I do know we're poised right on the brink of the harvest deluge. Tomatoes in both beds are ripening, sweet corn silks turning brown, five huge cucumbers hulking in the fridge, potato plants dying (meaning the spuds are ready to dig), peppers and eggplants beginning to shape up . . . Ted, the principal spaghetti sauce maker, looks at all this and groans. We always do groan, but it's what we've done everything else in order to achieve.

I cooked up some apples, mostly windfalls and all imperfect, to put through the food mill. We'll get a token amount of applesauce anyway.

August 6. Rob Cardillo from *Organic Gardening* came and took pictures, but didn't think them likely to be very successful. It was a bright day—sun too strong, shadows too dark, ducks too black and too jerky in their movements. All the same, his visit was diverting. I'm enjoying having other people come and look at the homestead in full feather, so to speak.

Our housecleaners, Diane and Donna, came today. Diane had asked a couple of months ago if she could watch while I opened a hive. The timing had always been wrong somehow, but today I wanted to work the hives anyway, so before they left I suited up and did the routine.

I'd put the two supers of extracted-out frames back on the second hive, for the bees to clean up at their convenience. Today the frames were immaculate, with just a little honey stored in one super. I left that super on the hive and took off the other to store. In the parent hive the medicated honey has mostly been eaten; that super wasn't heavy at all when I lifted it off. Underneath things appear to be pretty normal; I saw lots of capped brood and some healthy-seeming larvae, the laying pattern looks good, and there's no evident cause for concern. I didn't take the hive apart right down to its toenails, but will do that one day soon in preparation for combining the two hives.

In view of both hives' foraging behavior these past couple of days—field bees swooping up and away, no more hanging out on the porch all day—I conclude that something must have bloomed, and am taking a chance that the robbing of the parent hive by the swarm hive is over. I left the screened inner cover reversed, so the rear port would be inaccessible, but left the outer cover cocked open. And I changed the entrance reducer from small to medium. That should leave the hive able to defend itself, but give it adequate ventilation.

At dusk, when the bees had gone to bed, I mixed up a gallon of rotenone solution and hauled it out to spray the beans and cukes. Turns out they didn't need a whole gallon. What to do with the residue? I left it out in the garden overnight, knowing it will have lost force by tomorrow and can safely be dumped out. I sprayed all the bush cukes but only the bigger of the two Jazzers; I still want to see if the beetles leave that variety alone. No point in losing both plants if they don't, though. Poking the nozzle between brussels-sprout plants to spray the bush cukes, I noticed how chewed the upper leaves of the sprouts looked, and peered into their centers to discover a lot of small corn earworms chomping busily away. Not only rotenone then but, tomorrow, Bt.

Rabbits playing in the grass at teatime, birds still taking care of the imported cabbageworms. The water in the pond is crystal clear, thanks to the water hyacinths. One of the new ones Liz brought me on Sunday has bloomed—produced a stalk covered with delicate lavender flowers. Some weed! Yesterday a catbird went into the covered part of the duck pen for water or uneaten feed and couldn't find its way out. I "herded" it through the gate as I do the ducks.

August 7. A wonderful day, perfect weather. By 8:30 I had waded into the garden through heavy dew. There I spent two pleasant, unhurried hours planting fall lettuce and a whole bed of fall carrots, and erecting a duck barrier to keep the seven samurai from rooting up all the seed. Using the caged tomato row as part of the barrier, I needed only to prop two pieces of blueberry-arch fencing across the ends to block off the whole back garden. The arrangement is so makeshift that I wondered if the ducks would respect it, but they do; they've never pushed against any of their fences in an attempt to get out. The trick is more often to get them to *come* out into the big, dangerous world.

The rotenone at the recommended concentration—four tablespoons per gallon—burned some of the cuke leaves in a way I remember from two years back. Doesn't make me any happier about using the stuff. (This summer I've been directing an independent-study project in classic science-fiction and horror films of the Fifties, for a graduate student named Mark Shanaman; and watching *The Creature from the Black Lagoon* tonight, in preparation for Monday's session with Mark, I was amused to discover that rotenone is what they used to drug the Creature!)

This evening I poured out what was left, rinsed out the sprayer, mixed up some Bt for the brussels sprouts and cabbages, and lugged the refilled sprayer back to the garden. This is the first time all summer I've had to break out the Bt—*Bacillus thurengiensis*, a murky green fluid full of microorganisms that paralyze the gut muscles of insect larvae—but those corn earworms are murder, much more prolific and voracious than imported cabbageworms, or so it seems to me. And more interested in brassicas than in corn, if my experience is anything to go by: I saw no frass on the corn ears, not that that necessarily means the corn gets off scot-free. And alas, the birds—which gobble up imported cabbageworms as fast as they hatch out—don't seem to care for corn earworms any more than I do.

Three ducks were swimming today, while a fourth demonstrated that it could climb the ramp. The water in the tub is filthy. Once a week is none too often to change it, likewise the straw bedding, but more for our sake than for theirs—the first thing they do with a bucket of clean water is put mud in it. Yesterday the bucket I brought them when Rob, the photographer, came was opaque by the time he finished shooting. They scamper and chase around in their big run, apparently as happy as seven large black larks.

I sat beside the pond at lunchtime and fed the fish, first their shrimp pellets, which they treat like meat-and-potatoes and stop taking when they've had enough, then their flakes, which is dessert, and which they'd go on eating all day. They strike like a school of sharks when I begin; I was scooping out walnut leaves, and one struck at my finger! Spotted one of the catfish sidewinding across the bottom, also a tadpole, one with legs (small ones).

August 8. We drove up to Kutztown for the Rodale Research Center's annual event they call GardenFest!, which we've been attending for the past three years and always enjoy. Most interesting to me was the seminar run by Mary Appelhof, author of *Worms Eat My Garbage*, about how to set up a worm bin. Nothing to it; we've got everything we need but the redworms, which I'll order this week without fail.

A new company called NovaWood was exhibiting compost bins and park benches made out of recycled plastic. On impulse we bought their bench floor model. It was pricey, but buying a cheaper one would have required a trip to the store to get the molded plastic pieces and the lumber, plus an hour or two to measure and cut everything, and at this point in the season neither one of us has any business taking on new projects, period. However sensible.

So we sprang for it. Jeff Ball, who as usual was a featured GardenFest! speaker, kindly offered to bring the bench home in his Explorer, saving us the freight. My view is, two weeks from now we won't fret about the $180, but we'll be really glad we have something to sit on while contemplating the pond. Something that *didn't* have to be shopped for and assembled. Something, moreover, which is virtually indestructible, and virtuous to boot.

I remembered to bring my mystery bug, stashed in the freezer a month ago, but all day I failed to connect with the entomologist, Diane Matthews-Gehringer, who asked me two years ago to mail her a specimen. A quarter-inch beetle with wing covers like a striped cucumber beetle, but with a red head and underbody, found chiefly on potatoes. Makes a disgusting humped yellow larva that retains all its molts on its back. I left my specimen with somebody at the office. Next week I'll call to see if she knows what it is.

August 9. This was the day set aside in advance to finish pond framing. It involved a trip to Hechinger's for two eight-foot one-by-twelves and sixteen brass screws, and the project was frustrated by the loss of the wrench for the power drill. The four screws we planned to put in each corner became two when they had to be screwed in by hand. But the finished product looks great. The design is sandboxlike, a rectangular frame tied at the corners by one-by-twelves cut into equilateral trapezoids and screwed diagonally across the joins of the side and end boards. A design both simple and elegant, says Ted. And perfectly sturdy; I walked all the way around to test it, and you no longer have to be careful not to put any weight on the inside edges of the boards. So we're finally *really* done building; all that remains will be cleaning up the site, leveling it somewhat, and persuading some grass to grow there.

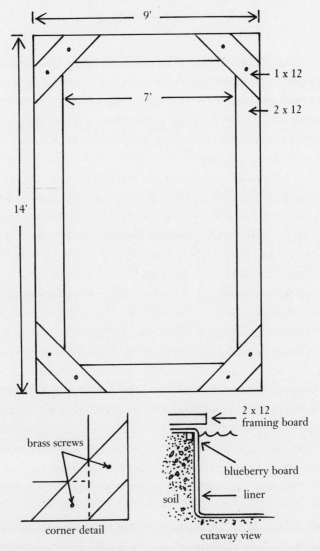

9'

1 x 12

2 x 12

7'

14'

brass screws

corner detail

2 x 12
framing board

blueberry board

soil

liner

cutaway view

I did the weekend duckmuck—*very* smelly this time—and siphoned the tub onto a different patch of grass. Then I pulled out the waterer and retired it from service. The ducks are big enough now to drink out of a bucket. When I emptied the waterer, I saw why it had become impossible to get the water cleaner by putting in more with a hose from the top. The whole central cylinder had filled with silt that smelled exactly like the silt at the margin of a freshwater pond: vile! The bacterial life in that water must have been rich beyond description.

The ducks had excavated a pothole in the mud next to the concrete apron, where the waterer was. I filled it with leftover clay, then put a window screen—the one I was using in the improvised solar wax melter—right across the muck and set the water bucket on that. They can't drill through the screening; maybe this arrangement will keep things a bit more sanitary.

They'd been put in the garden while I was seeing to all this. After finishing, I got a pan and went out to pick some beans. And caught the varmints red-handed, eating buds and leaves off the pepper plants and pecking at the little peppers; I could smell green-pepper smell from twenty feet away. I tried to chase them off, but the minute my back was turned they were at it again. So I guess they can't be turned loose in the garden anymore unless I can figure a way to keep them out of the peppers and eggplants.

I harvested Lars-Håkan's first cabbage, the only one I still know for certain was planted by him, for slaw, and finished picking wax beans and Kentucky Wonders. When it came time to herd the ducks back to their pen they were glad to go; but when they caught sight of the screen with the bucket sitting on it, and no waterer where the waterer had always been, they were dismayed. First they wouldn't go in. Then they stood and stared, struck dumb. Not until Ted and I had gone away to sit on the new bench by the pond did we see them finally venture to stand on the screening and drink from the bucket (and vainly try to drill in the overspill).

Beans and a big tomato for dinner, plus another box of wax beans frozen. Yesterday, the summer's first BLT: out of this world!

Late afternoon after late afternoon the bees do their swarming practice, spiraling up in the air, boiling out of the hive, flying in a cluster before it. It never comes to anything; after half an hour, or less, they simply go back in. I've never read anywhere about this behavior and haven't had a chance to ask Jim, but it does seem like requeening would be a real good idea. If that's how they behave in August, what about April?

August 10. A pleasant, farmwife sort of day. When I went out with my little containers of pellets and flakes to feed the fish this morning, three ducks were in their tub and two more were standing on the duckdeck. Five ramp-climbers! Later I went out to hang a basket of laundry and saw *four* black heads drifting above the water. The fifth duck had nestled down on the deck; the other two were huddled disconsolately in the mud below.

When I took my sandwich out to the picnic table at lunchtime, the four were still swimming; but after a bit one swimmer suddenly began to splash and dive wildly, sending the other three frantically over the side to get away from her. She threw out wave after wave of water. So much water had already slopped over the edge onto the dirt that the mud problem was getting pretty serious. After this display of hysterics, I got another screen out of the garage and put it down below the tub— this to the intense sorrow of the two remaining nonswimmers particularly, who had been working off some of their frustration by making deeper and deeper mud holes in the overslash.

The fish are *bigger.* I saw one catfish undulate across the bottom muck and a tadpole motionless in the depths; but we have an unusual problem here. The water in our fishpond is *too clear;* we have too little algae! That's why we can see the catfish and tadpoles, despite the fact that they show up poorly against the yellow mud on the bottom. The sides of the liner are clean black, hardly anything there for a snail to make a living on, and the water is crystal clear right to the bottom. Having read in Logsdon that "bluegills thrive on algae," I realized that we've overcorrected for that early algae bloom. So yesterday, when we went over to Liz's to pick up the bench, I borrowed an old tin washtub to serve as a holding tank for my water-hyacinth population. I'm going to take most of them out of the pond and keep them in the tub, then add and subtract until we get the balance right.

The fish are growing because we've been feeding them, and will continue to grow as long as we go on feeding, but water this clear goes too far in the direction of "ornamental." They're also supposed to be rustling as much of their own grub as possible from what's available, like the ducks do. It's supposed to be a balanced system, everything interlinked with everything else. Taking out *all* the algae-supporting nutrients, then feeding the fish entirely with commercial feed (and mosquito larvae), just ain't the homestead way.

Tonight we canned seven pints of pickles (Jazzer cukes) and froze four pint-and-a-half-size boxes of applesauce. The apples were very faulty and many were too green, but I'm grateful to be getting any sauce at all. Canning takes getting used to all over again every year. Tonight I ran short of pickling liquid due to not packing the cuke slices tight enough in the jars, and had to cook up some more in a hurry; and I didn't allow enough time for the canner to reach a rolling boil. At the second canning session I'm always much more adroit and get much less flustered, but at the first I'm a hopeless klutz.

I picked half a basket of tomatoes, Beefmasters and Viva Italias, for Ted to make spaghetti sauce out of tomorrow. He has a love-hate relationship with sauce making—complains and carries on about it, but when I first told him two years ago that there wasn't going to be a garden in '91, he got *the* most outraged look on his face and demanded: "No *garden*? What're we gonna *eat*?"

I watched the forecast and decided to water both gardens, pleased as always to have the trickle system, which does the job so well with so little effort on my part (once it's set up).

By the way, that batch of coleslaw made from Lars-Håkan's cabbage is the best yet.

August 11. Bad forecast; this afternoon we had a terrific thunderstorm, causing a power outage and knocking the top bar of the cucumber trellis off its uprights. Luckily Ted was home boiling sauce and could rush right out and put it back up. (I was in town, getting soaked.) The cuke vines seem undamaged by their mishap. A couple of eight-foot popcorn stalks fell over, but weren't harmed.

There's a little smut in the popcorn, none in the sweet corn.

Ted filled the washtub and put in one of the water-hyacinth

clusters. Without the pond's thin soup of fish manure and plant debris to feed on they'll need fertilizer—maybe a nice forkful of used duck bedding?

Sauce making and slaw making trapped Ted in the kitchen from 1:00 to 6:00 and made him cranky. Only six pints of sauce to show for all that work. But we ate what was left over on spaghettini for dinner, and even he admitted how very nice it was to reencounter that fresh, happy taste.

August 12. Half an inch of rain in the gauge. I put most of the remaining water hyacinths into the holding tank, discovering as I did so where all the algae-nourishing nutrients have been going. Several hyacinths are reproducing, in the multiple clonelike fashion of strawberries: send out a runner and grow a baby at the end.

The Black-Seeded Simpson lettuce is up thickly and unscathed so far by slugs. Maybe the ducks have reduced the population to the point where growing lettuce from seed will be a snap for once.

The word is in from Rodale's entomologist, Diane Matthews-Gehringer: my mystery bug is a three-lined potato beetle ("the old-fashioned potato bug"), Latin name *Lema trilineata*. And that's excrement the larvae carry on their backs, not skins from previous molts. Diane thanked me for the specimen. I promised a better-looking one next year.

In last year's Christmas letter I announced that I'd be throwing a fiftieth birthday party for myself on August 30, that everyone should reserve the date, etc. I've never thrown a big party in my life, or even a small one, I don't even like parties much. But fifty is a big one, so I thought why not celebrate in a big way? I knew the homestead—still a gleam in my eye at that point—would be in full flower, or rather in full fruit, and thought people might like to see it. But as the time gets closer, it's been looking just about impossible to get a party organized by the end of August. Also, maybe more importantly, I've realized that I'd much rather the homestead be the focus of the occasion instead of me.

So I spent much of today making up and having copied some

invitations to what I've decided to call the Homestead Open House, and scheduled it for September 19. I'll mail them out over the next couple of days. There's a current stamp with a wood duck on it; that seems appropriate. Like Harlan Hubbard always said: you have to let people come and look.

Ted thinks the idea of having a big party is nuts, considering (1) my general dislike of such things, (2) how busy we still are just keeping our heads above water, and (3) that the fall semester will have started by September 19. All true; yet I seem to want to do this anyway.

Five thousand redworms have been ordered from Beatrice Farms in Dawson, Georgia. At $19.95 postpaid, theirs was the best price. I've also found a source for garden worms: Liz says they have gadzillions over at her place, and that I'm welcome to some. Now to get the bins set up, side by side.

Yesterday Ted saw a largish bug fall into the pond and be torn apart and devoured: the bluegills, doing their piranha number.

I decided the ducks weren't getting enough vitamins, now that they can't be left alone in the garden anymore, and let them out into the yard today. Picking a time when I had half an hour to spare (for the first time in weeks!), I carried a chair to a spot between them and the pond and sat with my wand across my knees, like the fairy-tale goose-girl, while they bustled and clucked and ate grass and mushrooms. One duck—I guess it's always the same one—is mad for mushrooms and, if he spots one while being herded to the garden, will break formation to dart over and snatch it on the run, then scurry back to join the flock.

For a while all was well. But when they'd worked their way past the white oak, conditioning took over and they all started running toward the garden. Thwarted by the closed gate, they wandered away; but every time they'd get any distance from the gate, the sight of it would trigger the same response of rushing toward it. In the process they were getting closer and closer to the hives. In the end I let them go in for a little while (personally standing guard over the peppers) before herding them back home.

August 13. Our first Odorata water lily opened four days ago, closed up at evening, opened for a second day, then disappeared underwater. Another bud is about to break the surface. One of the Comanches is doing fairly well, the other isn't. For the past several days the bees have been using the mat of water clover as a place to drink from—one more homestead part integrated into the whole.

Already the water in the pond is less clear and there's more algae on the liner sides. Those water hyacinths are powerhouses!

This morning I moved the cold frame out of the garden, where its two halves have been standing on end for a couple of years, and reconstituted them into a third incarnation (after cold frame and solar melter) as worm bins. I'd originally planned to use the boxed-bed lumber, but it does seem that at four feet by eight the bins would be unnecessarily big. The cold-frame boxes are the same shape but a quarter that size. Like the boxed-bed frames, however, they've rotted away from their nails and fall apart when lifted and carried. So I laid them across the garden cart and wheeled them, one at a time, over to the space between the screenhouse and the ex-Island's sole remaining azalea—a bare weedy scrap of ground, perfect for a worm farm. It's right next to the pond, handy for tossing worms to fish, and the azalea will provide afternoon shade.

After setting up the boxes and loosening several inches of soil inside them, I filled one with used duck bedding (for the redworms) and the other with equal parts duck straw and leaf mold (for the garden worms). It's supposed to rain this afternoon; that will wet things down well, and then I'll mix the loose soil well into both, but especially into the second. What I particularly want to know is whether something in our soil is killing the earthworms as fast as they move in. Lids need to be devised for both these bins. We are finally beginning to run a little low on scrap lumber, but something will occur to me by the time my redworms arrive from Dawson, Georgia.

I looked up corn smut in the *O.G. Encyclopedia* and read that infested ears and tassels should be cut out and burned before the smuts burst and release the virus-carrying spores. There was a besmutted tassel on one stalk and a besmutted ear on another; I cut both and tried to burn them in the fire circle, but on this entirely windless day, getting a fire going was uphill work. I finally wrapped the smuts in newspaper and threw the bundle in the trash.

You can see how the Pueblos and Navahos, those drylanders, could come to view corn pollen as sacred. Its aroma *is* heavenly. The bees must agree; despite the fact that no reproductive purpose is served for the corn, corn being a wind-pollinated plant, bees of every size and type are working in the popcorn tassels—not just my honeybees, but bumblebees, little wasps, every creature that appreciates pollen.

The ducks are off their feed again. I gave them breakfast around 8:30, and they let it lie and went back under their house, where they spent the whole morning. Probably the weather; I felt like spending the morning under the house myself.

I scrubbed up and put away both the plastic waterer and the big galvanized one. Next spring, if we're successful at getting one of the ducks to hatch a brood, the waterers will be all clean and ready to set up again; for the present, into the basement they go.

Still can't tell which are ducks and which are drakes. Some of them quack, that's all I know. I thought at first that only the drakes would get the iridescent wing coverts, but they all seem to be developing them. Sunday they'll be eight weeks old—almost big enough to eat, God help us.

August 16. The past three days have been soggy and cold, with a total of 1.25 inches of rain, not good for gardening nor for duckmucking either. I've been home alone, Ted and his daughter Alison having driven down to Cape May for the worst shore weekend of the summer. Mostly I stayed in and worked on the unbelievable backlog in my study, actually glad of being forced to it by the rain. Before settling down Saturday and Sunday to make and address open-house invitations, I built a fire in the fireplace out of the punkiest, poorest wood in the woodpile plus a little dry kindling from the wood cupboard. I tried not to burn much of Ted's precious hoard of stovewood, but needed to drive the chill out of the air.

It worked. I stamped—colored—licked—sang along with my records of those cowboy songs the ducks like—and spent a pleasant couple of afternoons.

The trellis fell down twice more yesterday, and I had a time getting it put back up without stepping in the cucumber bed. It needs to be tied around the uprights close to the top, so the bar can't pull out; but working with plants is so disagreeable in wet weather—for them and for me as well—that I put it off.

I put off also the chore of standing those ten-foot-tall popcorn stalks upright again. They came down in Friday's wind—my punishment for not having hilled them when they were younger. Ted says that when a storm did this to the field corn around New Hope when he was a boy, the farmers would just grind the whole crop—stalks, ears, and all—since it was destined to be fodder for their stock. But if the popcorn is ever to be fodder for us I'll have to prop it up again somehow. Very hard when everything's so wet and shouldn't be handled, when the smut spores are just waiting to spread, when the soil I'll need to use for hilling is yellow clay, miserably heavy to dig or spread when saturated. Maybe tomorrow.

This afternoon I let the ducks out, meaning to put them in the garden; but they raced excitedly about on the lawn, flapping their wings like mad, and didn't want to go in. What they did want was perfectly clear: to poke around in the wet grass and mulch between the patio and the Golden Delicious tree, always sidling toward the hives (which I wouldn't let them near). They clucked intensely and gave every indication of deep satisfaction, so I stood around for a while with my wand and let them do as they liked.

But I couldn't really afford to stay out there with them long; so after the thrilling edge wore off, I put them in the garden. Two of the idiots wouldn't go in and flapped off wildly along the fence in both directions, leaping in and out of the electric-fence line (though I doubt they got much of a shock, after so many days of heavy cloud); and while

I was chasing and cursing after them, the other five bustled back out and went back to probing at the lawn with their bills and chuckling in that happy deep-throated way, too excited this time to mean contentment. I herded them back in and shut the gate, and they spent a couple of hours out there, but I got the picture right enough about where they'd rather be. Maybe the garden's so picked-over at this point that the lawn offers better forage—or maybe the ways of a duck are fathomless to humankind. Maybe we could devise some sort of makeshift enclosure so they could be left unsupervised on the lawn, if they like it so much.

The Odorata's new bud was just under the surface and starting to open when this stretch of rainy weather hit. Turns out water lilies are fair-weather flowers; if the timing's wrong, they miss their chance and never do bloom. That's a loss of fifty percent of this season's water lilies so far, a great pity. (Flowers don't in general grow very well for me, God knows, but I guess I can't take the credit/blame for this particular failure.)

The Comanches are *not* doing well. Call to Lilypons pending.

August 17. Rain, rain, rain . . .

I feel very bad about the popcorn, down since Friday, badly broken and bent, and no chance to do anything about it as long as it goes on raining. My inspection had revealed that something like half the stalks seemed damaged. It had been *so tall*—eight to ten feet—soaring over my head and even Ted's, and too spindly, top-heavy, and shallow-rooted to stand up to the storm.

Late this afternoon, when it was obvious the rain was never going to stop falling, I put on my rainjacket and boots and went out anyway to muck out the ducks, way overdue for clean bedding, and siphon out their tub. The pen and everything inside it were absolutely filthy and stank to high heaven. I dragged around feeling very heavyhearted because of the popcorn—amazing how hard one takes these losses—raking the matted straw out from under the duckhouse while rain fell on me, the straw, the dirt cart . . .

. . . and the ducks. I'd stashed them in the garden, out from underfoot while I was working. You'd think they would groove on this weather, but they were cranky and contrary, did *not* want to go in the garden, had to be chased in. Two took off in opposite directions around the fence, floundered in and out of the electric wire, quacking like maniacs. By the time I finally got them into the garden I was fed up with all seven of the little crosspatches.

I assumed the fence wouldn't have much umph left after the past four sunless days, but grabbed the wire to test it and was knocked out of my socks by a terrific jolt. Obviously the battery's still charged up. I just read somewhere that you're supposed to build a little house for the battery and charger of an electric-fencing system, they're not supposed to be left out in the weather. *Now* they tell me.

Under the prevailing conditions, these chores, never exactly fun, were thoroughly unpleasant. I felt like a character from *The Emigrants* before they left Sweden. The smell was overpowering, I wouldn't blame the Purbricks one bit if they filed a complaint to the borough about us. As soon as the old straw went out to the manure pile, though, and the old water into the grass (which certainly didn't need it), things improved a whole lot, smellwise.

When I'd finished with the ducks I took the plastic basket out to the garden and cheerlessly set about harvesting: the rest of the unrotted cabbages, four heads; a jillion tomatoes (I know what Ted's going to spend tomorrow doing); a few jalapeños; and lots and lots of Jazzer and Bush Baby cucumbers. Harvesting is fun in nice weather when you're ready for it; this was dreary. All I could think about was how Ted was going to groan and carry on when he saw the tomatoes, and how I'd have to spend a whole day canning pickles. Lots of millipedes, sow bugs, and baby slugs in the cabbages; I'm glad to see the last of those, truth to tell.

But! I'm *finally* beginning to feel caught up with the backlog of outdoor and indoor work. For all their dreariness the past four sodden days have given me a chance, not only to get out the open-house invitations, but also to clear up the more general mess in my study. I *am* caught up, for the first time since we got back from Utah, a month ago!

I look sadly at my handiwork, my realized vision, seeing how it's all coming together and cresting and even breaking a little over the crest already, and wish I'd been able to enjoy the *process* more. Now that I'm caught up, I can shift into maintenance mode—a mode in which I *can* enjoy process (canning, duck-sitting, harvesting) and not just product (getting it over with)—but summer's almost over! In four more weeks I'll be teaching!

I browse through this record with feelings balanced between exhilaration and irony. Exhilaration because the story so engrosses and engages me. Irony because I'm enjoying on the page what I wasn't able to enjoy often enough in the actual doing, like a shutterbug who's only able to appreciate his vacation when looking at the pictures afterward—who was too busy *taking* the pictures at the time to enter fully into the experience.

The trips away from the homestead are responsible for this kettle of fish. They were important, but they interfered in essential ways with what I needed to be doing here, and even more with *how* I needed to be doing it.

By rights, I think, the homestead experiment should run for two full years, with the second given to continuing what took so much time and effort (and money) to set up in the first. Though in actual fact, without exactly realizing it, I've been preparing for this ever since we moved here. We created the big garden and planted the apple trees that first spring, added the blueberry hedge in the second, added the bees in the fourth. Only the fishpond and ducklings are new—and the worms, when they arrive. In retrospect it looks like a five-year plan all along, if but a semiconscious one.

The other morning Ted saw an enormous black bird walking down the neighbors' driveway. It was a crow, but for a split second— Omigod!—he was sure one of the ducks had got out somehow, so much did the crow resemble them as it waddled/swaggered along.

August 18. The rain finally seems to have stopped. Before I left for town this morning, Ted and I went out and did what we could to rescue the popcorn: cut out the fatally damaged stalks, carefully stood the survivors upright—very bent, some of them—and shoveled leaden wet clay around their bases. In the process I discovered more smut. One way or another, this crop seems fated to fail; but we've done what we can, belatedly, to save whatever can be saved at this point.

The ducks' moods seemed as much improved by the change in the weather as mine was. They appeared to like their clean water—the old water may have been too much even for them. Five ducks were in the tub at once, and while hilling the popcorn we glanced over just as a *sixth* duck stepped into what little space remained. For the first time? Six ducks in one bathtub add up to one crammed bathtub.

When I got back from town, Ted had collapsed and ten pint freezer boxes of spaghetti sauce were lined up on the dining-room table. For dinner we had—surprise—spaghetti. The sauce is *great*—better (stronger, thicker) than the last batch, with surprisingly good frozen pepper bits from two years ago tossed in.

August 19. Another nice (though exhausting) farmwifely day. This morning I tied the flopped-over pepper plants to stakes cut from the jungle, a long-overdue chore. Let's hope the plants feel stimulated to start pumping out peppers in time for some of the sauce making. Peppers always lag behind tomatoes, no matter what varieties I plant, so the early sauce has next to none but there's a lot left at season's end to chop and freeze.

That done, I sat in a chair and cut all those broken popcorn stalks into four-inch lengths with my pruners, to add them to the compost. Normally I'd allow the chipper to do this job, but Ted says it's jammed and has to be taken apart. In the process of cutting the stalks down to compostable size I noticed quasi-mature ears on a few of them. The grains in these ears are as yellow as Butterfruit grains, not multi-colored like Pretty Pops are supposed to be. Another popcorn disappointment? Another crop that hasn't worked out?

I picked up some windfall apples. Too few to bother setting up the Juice Mate, so I peeled, cored, and sliced them, a job that reminded me, as not too many jobs do, of the arthritis in my hands.

Canning pickles was the day's major chore. In the end the count stood at five quarts and two pints of whole dill pickles and dill spears and slices, plus four quarts of bread-and-butter pickles. I took the easy route and used Ball package mixes for both batches. When Ted got home I was as wiped out from *my* day on my feet in the kitchen as he'd been yesterday from his.

Observations: two more Bush Baby cuke plants are dead of wilt. Rather than spray again I'm inclined to just let them die if they've a mind to. Despite my earlier predictions we've surely got enough pickles now to last us all year.

The fish are *very definitely* bigger.

Twice, as I've approached the pond, I've seen the quick streak of something jumping in from the side. Has one of the tadpoles metamorphosed already? I wouldn't have thought so, but this behavior I'm almost-seeing certainly is froglike.

I keep forgetting to mention that all three types of lettuce and both types of carrots came up thickly; then we immediately had four soggy days in a row, after which there wasn't a single lettuce seedling to be seen and almost not a carrot seedling either. Zapped by slugs, dammit. And no way to fight back. If I'd let the ducks in there, they'd have eaten slugs and seedlings indiscriminately, and the results would be the same. Successful direct seeding of these small-seeded crops just wasn't possible this year; too many slugs survived the earlier depredations of the ducks. It would probably make better sense now to fill those carefully prepared beds with *flowers*: asters and mums, a little seasonal color for a change.

During a break I read up on butchering ducks. It's true: right after their first complete suit of feathers comes in they go into a heavy molt, during which time "the flesh is not in prime condition." We really ought to slaughter at least two of them pretty quick. They seem to be fully feathered now; the last little roughnesses of down are gone even from their necks. Oh dear.

All of a sudden the sweet corn's *ready*. Tonight's dinner menu: corn on the cob, tomatoes, coleslaw with onions—all ours. Beans would have given us a complete protein, but I just didn't feel up to bothering somehow. What we had, anyway, was complete *homestead*.

August 20. Ted was right—lids for the worm bins *were* discoverable among the scrap and salvage. That old kitchen counter, saved for the greenhouse, exactly covers the back halves of both bins, and two pieces of plywood exactly cover the front halves. We're now completely set up for business, lacking only the worms.

Today I took some time, just for a change, to sit on the bench and watch the pond. I counted not twelve but either fourteen or fifteen bluegills. Some of them swim into the Anacharis foliage and so miss the count; but if you watch for a while they swim out again. Saw also one of the bullfrog tadpoles (back legs but no front ones), a catfish, drinking bees on the water clover, just the right amount of algae on the lily-pad stems, and—best and most unexpected of all—two frogs! Two! One bigger than the other, but both clearly leopard frogs, not metamor-phosed bullfrogs. Where can they have come from?? Not wandering overland, surely . . . but where else? I saw first the larger, floating with just its eyes above the surface, exactly like frogs are supposed to do, and then the smaller surfacing nearby. Both were in the warm, shallow water by the shelf, where there's a space vacated by the defunct cattail plus lots of cover. Neither frog caught a bug while I was watching, but I didn't watch too long; Ted came home and walked over, and my calling to him to come slowly made them both duck under. But they came up again before long, so he got to see them too. The big frog floated for a while with its face in the notch of an Odorata lily pad. A pale blue dragonfly— very small, but not a damselfly—flew in and out of this charming pic-ture.

Marvelous sensation of the various homestead elements com-ing together into one thing, a system, Gaia in miniature.

Suddenly the corn is pouring in. We ate some again tonight, froze six nearly perfect ears (without blanching) to eat right away, like next week, and the rest, mostly pretty blemished by those tiny worms that turn individual kernels brown, got blanched and cut off the cob to freeze as cut corn. Don't know what the worms are; the description of seed-corn maggots fits, but those are supposed to attack the seeds you plant, not the ones on the ear about to be harvested.

August 21. A stunningly beautiful day. Vicki's kids were coming over to see the ducks, so I siphoned the pond in the morning; only four days after being cleaned it was absolutely filthy, and I didn't want Amanda and Laura to be grossed out. There were feathers in the water and feathers on the ground. Impossible to ignore it: the molt is starting. We're going to have to butcher two of them. When Vicki and I drove over to Liz's to get a bucketful of worms, Liz agreed to photograph both the slaughtering and the butchering—an indication of her professional zeal, because no more than my sister can she imagine how we're going to kill creatures we've raised from babies and fussed and chuckled over as we have these ducks.

We'll kill two now, then two more in six weeks or so—but *after* the open house, so there'll still be five ducks for guests to see.

Cartoon: gray eminence in knee breeches, with staff, says to stubborn-looking duck wearing kerchief: " 'But they are dead, these two are dead! / Their spirits are in heaven!' / 'Twas throwing words away, for still / The little duck would have her will, / And say: 'Nay, we are seven.' "

The baby pea vines in Liz's garden inspired me to emulate her example. I'd been thinking maybe spinach where the lettuce and carrots met their doom, but spinach comes up little—i.e., slug-vulnerable, too. But peas! Those nice big seeds make such robust seedlings, much less likely to be wiped out by slugs. What's more, I've got some snow-pea seed from '90 in the freezer. Make it so.

I added a shovelful of pond-excavation topsoil to the left-hand worm bin, already filled with duck bedding and leaf mold, and poured in Liz's worms.

Vicki's husband, Chris, arrived with the kids around 5:30, and right away I suited up and went to open the parent hive, since that was the item of planned entertainment they'd been promised. And there I made a dismaying discovery. The instant I set the smoker down I smelled the rot, and one look told me what had been going on. The carpet under that hive was adrift in dead bees. The swarm hive has obviously *not* given up robbing. Some of the weird activity I've observed must have been that, and not swarming practice or flying practice or a collective nervous breakdown. Bees from both hives must have died in great numbers. There they all were, heaped up and smelling fearful.

I went ahead and opened the parent hive, expecting to find signs of serious decline; but instead there were great sheets of capped brood and a fair amount of uncapped brood—and a full hive body of honey, in addition to a full super. Some of the wax was ripped off jaggedly, the way they say robber bees will open honey cells in the enemy hive. I didn't see the queen; but there's nothing wrong in there, except that the swarm hive won't stop robbing it. Time and past time to take action.

Amanda, ten, and Laura, seven, had a good time feeding bread to the ducks and fish flakes to the fish. They were fairly interested in the garden, too. "Can we have a homestead?" said Laura hopefully to her mother after we'd driven the ducks into the garden and back again. Vicki's main observation was that until today she hadn't heard me talk about what increasingly preoccupies me: the *interconnectedness* of everything. Besides making honey, the bees pollinate the garden and fruit crops, while the garden provides them with pollen and nectar and the pond provides them with water; the ducks eat slugs and bugs (and cabbage and bolted lettuce), their manure goes into the compost, worm bins, and water-hyacinth nursery, their tub water full of dissolved manure is siphoned into the lawn, and they'll ultimately provide us with meat and eggs; the manure-nourished grass clippings mulch the garden beds and feed the tomatoes, etc.; the worms compost household garbage and feed the fish; the fish will feed us.

I contemplate all this through Vicki's acute, appreciative eyes and glow with pleasure and satisfaction. Vicki's my best and oldest friend at Penn—indeed my only close friend there, Ted always excepted. In many ways I've never been closer to anyone, but homesteading at this level isn't in Vicki's line, and it's been a little sad for both of us that she couldn't share more fully in an experience that has so absorbed me—just as it's been sad all along that the literary criticism on which Vicki's international reputation as a leading Joycean is founded isn't in *my* line. But nobody else could see more perfectly what it means to me to have brought this project off.

August 22. While Ted was sifting compost and cutting the grass this morning, I sprayed Bt on the brussels sprouts and cabbage stumps (some of which are growing little auxiliary heads); tied a surcingle around the trellis so it won't fall down again; planted peas where lettuce and carrots were intended to grow; and picked a full basket of tomatoes—more than I thought were out there, far more than those dilapidated vines look capable of producing. The peas: Oregon Sugar Pod II, a tall, productive snow pea, excellent in stir-fries with onions and pieces of marinated chicken (or duck?).

And I thought out what to do about the bees. I first noticed the robbing on the day we extracted honey, two weeks ago. There's been abnormal activity over there ever since, and while I think some of it really was that weird quasi-swarming behavior, some of it was pretty certainly more robbing.

There doesn't seem to be much of a flow on, though field bees from both hives have been coming in with big wads of bright yellow pollen—corn pollen, most likely. In the absence of a honeyflow, I don't know how robbing can ever be stopped once it's well-established behavior. So the best solution really does seem to be turning the old newspaper trick once again: put the weaker hive on top of the stronger and eliminate the adversarial relationship by uniting both colonies into one big happy family. I'd meant to do that anyway this fall, though the parent colony looks to be in a lot better shape then expected, with plenty of honey and oodles of broodles, and I might think again if the robbing weren't a factor.

But it is. So I thought this out as a two-step plan, and after lunch

I suited up to set about Step 1: reducing the two-story parent hive to a single box by putting all the frames containing any brood at all, from both boxes, into the lower box, and all with honey but no brood into the upper. The lower box I left on the stand, with its screened inner cover on top. The upper box, with its super of medicated honey on top, I left in the grass with the telescoping cover on tight. I shook and brushed as many bees off the frames of honey as I could, but shut the rest of them in for the remainder of the day, in order to keep bees from both hives from robbing out the honey.

I'd left a second super on the swarm hive, in case the bees produced some fall honey (from goldenrod), but there was very little honey and none of it was capped. What I found there had probably been purloined from the parent hive anyway. I took that super off too and set it aside, for the bees from both hives to clean out. So the parent hive was one story tall, and the swarm hive two stories plus one super tall, for most of today.

Ted's elder son, Sandy, and his family stopped by unexpectedly in the midst of these operations. It was bad luck that I'd tackled the problem on a day when four kids under twelve were going to be running around the yard with bare feet. The bees were terribly agitated, guard bees flying all over the place. Both Sandy's wife, Jade, and I got bees tangled in our hair. Neither of us was stung, nor were any of the barefoot kids, but that was pure luck.

Before dinner I went back and took the cover off the honey-filled hive body and super from the parent hive, so the trapped bees could go home for the night. I wanted as many bees as possible to be in the bottom of the parent hive before I moved it; there won't be any way for them to get in afterwards, even if they knew enough to try.

After dinner, with barely enough light to see, I went back. Almost all activity had ceased, almost all the bees had vacated the box and super full of honey. I uncovered the swarm hive and laid some newspapers on top—I'd poked toothpick holes in *two* sheets of newspaper; I want those bees to take a *long* time chewing through. The inner cover

COMBINING TWO HIVES INTO ONE

STEP ONE

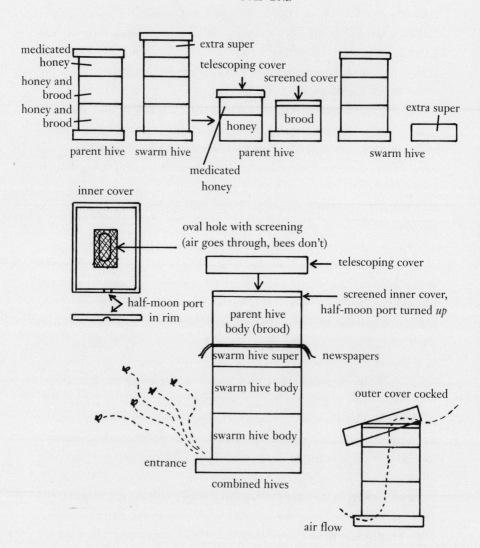

STEP TWO

was solid bees underneath, the exposed tops of the frames likewise. They were quiet and unaggressive, which was good because I wasn't using smoke.

When the papers were on, and creased slightly to hold their position, I pried the parent hive body full of brood, together with its screened cover, off the bottom board, and set it atop the newspapers. Then I offset the telescoping cover on top of the whole thing—offset to provide ventilation but not access. The port was turned up and the oval hole, of course, was screened. I want there to be no way out but through the newspaper-blocked lower route.

I left the top box and super of honey from the parent hive, both still containing some bees, askew atop each other in the grass with the cover off, and set the cleaned-out super from the swarm hive onto the bottom board of the parent hive, to give these stragglers someplace to go. Then I left them to get on with it and came in, bringing several angry guard bees with me as far as the light above the door, which interested them more than I did.

Finally, after dark, I went out for the last time with the cart to bring the honey in. The super I took to the basement; the bees will get it back as soon as they've settled down together. Of the nine deep frames from the parent hive, four are full on both sides with capped honey. Into the freezer they go; I'll feed that honey back to the hive, instead of sugar syrup, in the late winter. The other five deep frames were only partially filled; those I carted back for the bees to clean out in the morning.

Most of the bees were gone from these two boxes, but not all; so before I could bring the frames full of honey into the house, I had to stand on the patio and carefully brush bees off each frame before setting it inside on spread newspapers. Four deeps and nine shallows. Phew.

Visitors or no visitors, we also had to process the tomatoes this afternoon; some of them had rotten places and many were very ripe. The kids pitched in like gangbusters. I cut up tomatoes in the kitchen, Katy relayed the pieces into the dining room where Ted had set up the Juice Mate ("I'm the messenger," she proudly told the camcorder),

Christo manned the wooden pestle and/or turned the crank, Charlie wielded the spatula, and the whole job went like lightning, though this was the biggest batch of sauce yet—twelve pint boxes, and prime quality. For dinner: spaghetti. For dessert: cooked windfall apples, with honey, cinnamon, nutmeg, and whipped cream. (Lunch, by the way, was tuna-salad sandwiches with some of the sliced pickles chopped in. Good flavor, but wet and soft. How to get a home-canned pickle to stay crunchy?)

August 23. An anxious, trying day. We'd planned to butcher a duck this morning, but during breakfast I noticed a lot of bee activity outside. When we went to investigate we found the whole yard, front and back, buzzing with riled-up bees, and a full-scale frenzy going on over by the exposed frames of honey in the cart.

Ulp. So now I know better than to leave honey out when the bees are already upset; but there was no way to halt the furor already in progress. I called the Stipes to warn them to be careful if they went outside. Helen had already noticed lots of bees on her terrace and wasn't surprised to hear my voice on the phone. I took over some honey later; I'd meant to anyway, and this seemed an especially good time. She was understanding, luckily, and hadn't planned any yard activities for the day.

So the duck project had to be postponed. I wasn't sure whether to feel relieved or not, since I still had to go through with it sooner or later. And in fact it was sooner. By lunchtime the bees had completely cleaned out the frames and calmed down. And around 1:30 we trudged out gloomily to do the deed.

I'd hoped the ducks would be swimming so I could pick out a nonswimmer. But as bad luck would have it they were enjoying a siesta under their house, and I had to spook them out with a wand in order to choose a victim. Under the circumstances there wasn't much to choose between. I drove the lot of them into the nonbathtub end of the run and grabbed a loudly quacking, reasonably plump-seeming duck by the neck with my right hand. Then I picked it up as the books instruct: by sliding my left hand under its chest, grabbing its legs, and holding it so it was sitting on my left arm, legs secured by my left hand and wings by my right.

It sat there quite comfortably, quacking loudly at first, then just looking about while I carried it toward the gallows Ted had constructed: a length of wood lashed across one of the compost bins.

That was the hardest moment for me. Birds have a higher body temperature than humans do, and the duck was so *warm*. The last thing in the world I wanted to do was help kill that warm little body; yet I gritted my teeth and did it anyway. Ted tried to tie its feet while I held it, but the noose slipped loose, so I gave the duck to him to hold, tied its feet myself, and held the rope taut when Ted let go so the duck was hanging upside down.

Once caught, the duck was amazingly calm about all this, even about being hung by its feet; but I wasn't a bit calm and neither was Ted. He had a stick ready to clout it with, and a knife to cut the throat when he'd stunned it. The stunning went well, but the knife was too small and the stunned duck's throat too well feathered and too floppy; the first cuts he made were shallow, and the duck started to come to and struggle. The second blow may have killed it, certainly knocked it out, and finally Ted found the right place—behind the bill on the left (our right)—and blood started spurting. And soon after that it was over. The whole thing, mistakes and all, only took a couple of minutes.

They were certainly long minutes for me. I tell myself we were a lot quicker than a fox, which wouldn't have bothered to stun the duck first, or a raccoon, which would have bitten out its crop—both more natural fates than being hung, clouted, cut, and bled. But I wish, as always, that the system here on Earth didn't require some creatures to die in pain and fear so that others can live.

There isn't very much blood in a little duck. In a minute the spurting had stopped, and we took it down—still disturbingly warm—and carried it out of the garden by the feet.

We'd moved the picnic table farther away from the duck pen; it had struck us both that to skin and eviscerate this duck right smack in front of the others would be callous. I spread newspapers—where would this homestead be without newspapers?—and began, Ted reading out the skinning directions from *Raising the Home Duck Flock*, the most all-around helpful of my three duck books. First, cut off the head and feet with the cleaver: Clop. Clop. Clop. Then the last joint of the wings: Clip. Clip. Then take a sharp knife and slit the skin from neck to vent, making cuts around both sides of the vent. I did that—with some difficulty; the knife I was using wasn't sharp enough, and a quick search turned up no others much better for the job than that one.

Having made the slit more or less as directed, I started removing the skin, "which requires a good deal of pulling," says my trusty guidebook (an understatement).

I wonder if plucking wouldn't be easier. At least, the feathers came out in handfuls, even without scalding, and the skin stuck tight. Skinning a plucked chicken is one thing; pulling the skin off a bird with the feathers still on is another. I managed it, but, between emotional and physical stress, got really worn out. Ted brought me a bucket of water for rinsing my hands, and a plastic bag (from Eddie Bauer!) for the head, feet, wing tips, and hanks of skin and feathers, as these were separated from the carcass one by one. I didn't try to save the feathers and down, not having thought of any use for them, and having as much to do as I could handle anyway. I tugged and yanked, my hands getting sticky and down-covered—I had to rinse them repeatedly in the bucket—and finally the skin was all off except for the bit on the pope's nose, where I just couldn't seem to get a grip with knife or fingers. Finally I used the cleaver one last time to chop the whole thing off, oil glands and all.

Whew. Ted fetched two bowls, and we proceeded straight to the next step: evisceration. Compared to skinning, eviscerating was a snap; I knew just how to proceed, thanks to my college bio minor. First you make a slit between breastbone and vent, not cutting the vent or the intestines lying just behind the skin, and pull out the viscera. These emerged in a neat bundle: small pink coil of intestines, large tough gizzard, liver in two lobes, with the green gallbladder attached (I tried not to break it but failed, courtesy of the dull knife). I had to reach way up inside to get the heart out. Trachea and esophagus came out together. I didn't remove the lungs, which were embedded in the back and tore when pulled.

Intestines, etc. went into the discard bowl, giblets into another bowl along with the neck (one more chop of the cleaver). The gut was almost empty, though the ducks were fed this morning after I realized we weren't going to be able to start butchering till the bees settled down; either this one didn't eat much or food whips through them in no time flat. I slit the gizzard to rinse out its contents: sand from the grit bowl, but no nails or shreds of glass or diamond rings. The bowl of giblets looked *exactly* like miniature versions of what you find inside the Thanksgiving turkey from the supermarket. The carcass itself—all red meat and unbelievably small without its feathers—looked more like a skinned rabbit than a turkey.

I took this object inside, washed it well, rubbed vegetable oil all over it so it wouldn't dry out, and put it into a big Ziploc freezer bag to age in the refrigerator for "twelve to thirty-six hours." The giblets went into a little Ziploc bag, and then I collapsed in the La-Z-Boy chair, too exhausted even to help bring in and dispose of the newspapers and offal.

The day had already been traumatic enough, but now I had to think ahead, because several things needed to be done with the violently agitated bees. The hive body with the five bee-emptied frames had to be brought back for cleaning and storage; and the super of stragglers, left overnight on the parent hive stand, had to be put on top of the combined hive, over an inner cover with a bee escape in the oval hole. The fleeting thought occurred to me that it would be better to put together the new inner cover I'd bought but never built, because it has no half-moon port in the rim. But the thought vanished until later events gave me cause to remember.

I did go back at almost-dark to do these chores. All would have been well, except that the bees in the top hive body, confined there for twenty-four hours by the newspapers and separated from their honey, were *furious*. I set the super of stragglers atop the stack; but when I cracked the screened inner cover loose from the hive body, to replace it with the other one containing the bee escape, they came boiling through the crack.

I should have stopped then and there and let the stragglers perish. But I went ahead and pushed the cover with the bee escape between the hive and the super of stragglers, pushing the screened inner cover off as I did so. And, of course, the new cover had a port in the rim. So the bees could get out; and while I was trying to plug the port, with leaves or grass or *something*, they stung my right wrist three times through the gauntlet.

But then they started going back in through the port—they realized what time of day it was, I guess—and I went off with the emptied hive body to inspect my stings. The super of stragglers should be cleared out by this time tomorrow night, and I can put the screened cover back on.

August 24. I thought it might be interesting to record what I actually did all day today, a fairly average farm-wife day in most respects. So:

Got up and had breakfast, then fed the ducks, who weren't very interested in eating—maybe the molt (or grief?) has put them off their feed. Did my back exercises and got dressed. Folded and put away laundry from last week, made the bed, cleaned up the kitchen. Husked corn picked yesterday—this supersweet Butterfruit doesn't convert to starch, so you can get away with leaving it overnight—cutting out the bad spots, sorting the ears into *dinner* and *to be cut off cob and frozen*, and packing them into two plastic bags. Took out the compost bowl heaped high with corn shucks, and a plate of cut (faulty) corn for the ducks, who appeared indifferent to it. Dug the garbage into the bin, adding rotten tomatoes from the vines. Noted with approval that Ted had stirred around the more-advanced compost in the other bin to cover up the duck blood.

In the garden I noticed two more good-size eggplants, and that I'll need to water the peas if it doesn't rain soon; they need to get going. The corn is about done. Probably spacing was too tight, especially in the triangle. I doubt we got a bigger harvest than usual—impossible to compare, with so many partially filled-out ears. Rodale, where I got the idea of hill planting under plastic, isn't growing corn that way this year, but of course they've run that experiment and are trying something else now.

I dumped, cleaned, and refilled the duckbucket, hosed the duckmuck out of the bottom of their tub, which I'd siphoned last night, hung in the hose, and let the tub fill. The ducks were still ignoring the plate of corn, but several seemed intensely interested in the filling tub. While the water ran in, I scraped the hive body I'd taken off the parent hive, whose edges were caked with propolis and squashed bees. I also scraped propolis and dead bees from the five frames the frenzied bees had cleaned out the day before. Curious bees kept bugging me as I worked, smelling the strong, sweet fragrance of the comb, so I put the hive body in the house and added the frames one by one as I cleaned them. The fragrance of the hive is so intensely good that the stench of rotting deader, a few days back, was a special shock. One day last week I worked between the two delights of corn pollen and hive fragrance, each more wonderful than the other.

Between frames I kept checking to see if the tub was full; several times I've gotten busy with some project and let it overflow, adding to the duckmuck problems in the pen. But this time I caught it right and turned the water off. Two of the ducks jumped in when the water was still a foot below the deck board, and two more joined them as the level rose. I think the molt must be itchy and that swimming relieves the itchiness, but who knows.

When I'd cleaned up all the frames, the box went into the back room for storage. I don't expect to need it again till swarming-time next spring. At the hive the bees seemed calm enough, some flying from the rear port. No sign yet of any newspaper crumbs at the entrance to indicate they're chewing through the barrier.

I consulted a couple of cookbooks for directions about cooking duck and making cucumber salad, and made a note to pick up some vinegar, wondering briefly if it wouldn't be possible for us to make cider vinegar from our apples. Tonight's dinner is to be one-hundred-percent homestead-grown. Our chief unmet need is a source of milk: yo-gurt for salad dressings, milk for milk, butter (occasionally). Solution: the goat I realized I couldn't get away with keeping in suburbia. I weighed the duck together with its giblets: two pounds five ounces. The books suggest broiling or frying for ducks under three pounds; I guess we'll broil ours.

Lunchtime. A sandwich made mostly of tomatoes. The ripest one in the windowsill was so perfect that I decided to save it for din-ner, and went out to the garden for another.

After lunch I made a double batch of bran muffins, my breakfast staple. When they came out of the oven, I went and picked all the ripe tomatoes and again hauled an almost-full basket back to the house, realiz-ing Ted would have a fit when he came home and saw it. (Indeed.) I took a bucket out to check on the apples, but the Prima is pretty much bare and the Liberty doesn't appear to be quite ready—even the red-cheeked apples are very green. Wait a week or so, then. The two lone Golden Delicious apples still on the tree are flawless but green also—but I re-member last year they were still green even when exquisitely ripe. Soon.

These ducks just don't *like* raw corn.

The day was so beautiful I decided to go for a walk before run-ning errands in Media (take *Enemy Mine* back to the video store, buy vinegar and milk). That occupied me up to teatime. After tea I cleaned up the kitchen. Bill called to say he and Jeff would be stopping by in twenty minutes to check on the leak in the bathroom sink. I hung up, stepped outside—and saw to my appalled amazement that the bees were in robbing mode again, an intense mob of them bobbing up and down before the entrance of the hive.

Who was fighting who? I couldn't see, but things looked bad. When I hurried into my bee suit and went over to check, I was more appalled to realize that by leaving an entry open above the upper hive body, the one housing the former parent hive, I'd made it possible for the bees from that hive—with all their honey stores gone—to come around and try to rob the bottom hive, turning the tables neatly: the former robbers had become the victims of robbing.

Either way, they're just as dead. Heaps of the freshly dead lay on the grass. It was awful, I felt sick. Everything I do to try to help them stop killing each other seems to result in more killing, through my own inexperience, ineptitude, and misjudgment.

The only thing I could think of to do was take the entrance reducer from the empty stand and stick it in, to make it easier for the swarm hive to defend itself. So I did that, and in the process got my fourth sting in two days, this one in my upper right arm, through the coverall and my shirtsleeve. Not a bad sting, but it hurt, and the surprise of it threw me off balance. You can't let this nylon coverall fabric stretch tight against you and expect it to afford much protection. One of the three stings from last night has been bothersome today, hot and quite swollen, making it hard to bend my wrist; the other two didn't amount to much.

Bill and Jeff came as I was retreating from the bee situation (followed by furious guard bees, much farther than usual—they've been bothered for four days running, no wonder they're so mad). They had noticed there were only six ducks in the pen and guessed what had happened to the seventh. When I showed them the carcass in the fridge, Jeff marveled, as had I, at how little it looked, compared to the living feather-coated birds. I said we'd decided to let the others put on more weight and turn into roasters before doing this again.

While Bill and Jeff worked on the leaky basin, I worked at the computer. At 6:00 I started cooking: peeling a big cuke, setting out the duck. I quickly got overwhelmed, and Ted—who'd arrived home half an hour before—came out to help. He took over the duck project; I put my bee suit back on and went to cut the bee grass.

Since bees are deaf* they don't appear to mind the racket of the mower, usually. But this time some of them flew around angrily

---

*Or so it was believed in 1992. Recent experiments have established that "bees use a structure called the Johnston's organ, a chordotonal organ made up of nerve cells in the second joint of a bee's antennae, to pick up airborne sounds." Wolfgang H. Kirchner and William F. Towne, "The Sensory Basis of the Honeybee's Dance Language," *Scientific American*, 270 (June 1994), p. 79.

when I was cutting close to the hives. The mower threw up the terrible stench of the rotten corpses, but also chopped up and dispersed the newly dead; maybe the smell won't get so bad this time. When I'd finished mowing I removed the super from the hive, took off the cover with the bee escape, and put the screened cover back on. So once again the bees in the upper hive body have no way out except down through the newspaper. Get chewing, gals.

The bee escape worked well this time; the super was clear. Now I'll leave the poor things alone for a while, apart from checking in a day or two to be sure they got through the newspaper and none are still trapped in the top.

I lugged all the superfluous hive parts back to the patio. Then I pulled off my veil and gloves and went in to dinner, which by that time was almost ready. The menu: broiled duckling a la Homestead, corn on the cob, sliced tomatoes, cucumber salad, all homestead produce barring the various condiments, and every item excellent of its kind.

Even oiled and basted the skinless duckling dried up somewhat under the broiler, but still it was pretty good: mild-flavored, not at all rank or strong, and not a bit greasy. Our first experience of providing *meat* for our own table, an unsettling yet exhilarating feeling. I felt chiefly a great determination to cook the other ducks as successfully as possible, so as not to cheapen in any way the sacrifice of their lives to our necessity—something I never yet felt about a supermarket chicken or a pound of ground beef in plastic wrap. It seems an essential discovery.

In *A Country Year*, Sue Hubbell says this of her neighbors:

> These Ozarkers do not question the happy fact that they are at the top of the food chain, but kill to eat what swims in the rivers and walks in the woods, and accept as a matter of course that it takes life to maintain life. In this they are more responsible than I am; I buy my meat in neat sanitized packages from the grocery store.
>
> Troubled by this a few years back, I raised a dozen chickens as meat birds, then killed and dressed the lot, but found that killing chicken Number Twelve was no easier than killing chicken Number One. I didn't like taking responsibility for killing my own meat, and went back to buying

it at the grocery store. I concluded sourly that righteousness and consistency are not my strong points, since it bothered me not at all to pull a carrot from the garden, an act quite as life-ending as shooting a deer.

I see the situation exactly as she does, but I guess I feel just different enough about it to be able to do butchery—and imagine going on doing it—if circumstances permitted or required me to.

Dinner concluded, we blanched the rest of the corn and set the bones and giblets to simmer. No time this evening to freeze the stock or cut the corn off the cobs and freeze that; it was getting on toward 9:00 and I still had to screen *On the Beach*, a two-hour-plus movie, for my final session with my graduate student, Mark, tomorrow. Thus do the oddments and endments of one day get carried forward to the next.

We watched the movie, penned up the ducks, and turned in. Duck dinner excepted, a not atypical day down on the homestead.

Incidentally, this sort of overview shows me, as nothing else could, how thoroughly I've learned the lesson of how to work all day at a steady pace—fall into an easy rhythm when I start an outside chore, instead of rushing around wasting energy the way I did earlier on.

August 25. While I was in town today, Ted froze fourteen pints of thinnish (he always cooks it till he can't stand any more and then quits, however thick it is or isn't) but very delicious spaghetti sauce, which brings the tally to forty-three pints.

The bees have settled down again, apparently. This evening they're clinging to the front of the hive and the bottom board in another great beard. I'd dearly love to know if the beard is made up of bees from both former hives. In another day or two, I'll *have* to look in to be sure the newspaper has been penetrated and bees are moving up and down inside the hive. But it seems best to wait as long as I possibly can before disturbing them again.

August 26. I took the day off in order to try to contemplate the fact that I'm about to turn fifty. I've been concentrating so hard on homesteading that the past months have been spent less introspectively than any I can recall from recent years. Getting back into self-reflective mode, just like that, is hard to do. The closest I managed to come—sitting on the bench by the pond after feeding the fish, counting bumpy frog faces in the bog plants (now up to *five!*)—was a sense of deep satisfaction in what I've accomplished here with Ted's help, and a recognition that the homestead has been this year's major creative achievement. Sitting out there, gazing from garden to beehive to fishpond to duck pen, was like browsing through one of my own completed novels—still in manuscript, lots of work still to do before I'll hold a published book in my hands, but with the great creative effort completed and behind me.

It wasn't easy, but I managed to resist the impulse to do any work all day long, till dinnertime. Dinner was terrific. I'd simmered the giblets and the bones from the first duck dinner into a dark brown broth almost free of fat. We made Pepperidge Farm stovetop stuffing with some of that stock and used the rest to make giblet gravy. I pulled the meat off the breast and wings—all of it dark, strange to behold!—sliced it thin, and heated it in the gravy. We spooned the meat and gravy over the stuffing on two plates and served it with Ted's coleslaw, the last of the season, and it was *out of this world*—full justice done this time to the paschal duck.

Ted says he is sick and tired of being criticized and condemned by everybody when he tells them about the ducks. He's of a generation that can remember going to a farm and buying a chicken; the farmer would go out and catch one, wring its neck or chop off its head, and give it to the customer bloody stump and all, and the customer would take it home to pluck and cook. I can barely remember Mom's Aunt Bertha, the gentlest soul in Christendom, doing that once when I was a kid: stringing a chicken up by its feet and sawing off its head with a butcher knife.

The academics we know at Penn, the younger ones especially, have no context for that behavior, and consequently none for this of ours. I think Ted's unrealistic in his expectations. But being condemned by those who, unlike Sue Hubbell, haven't thought the issue through *is* tiresome. I suggested that he just not talk abut the ducks at Penn.

The one bit of work I did was to check the hive. I decided not to take the inner cover off the top, but to unstick the two hives at the newspaper line, slide the top hive body forward, and tilt it up to see if there were holes in the paper. This turned out to be easier than I thought, because while the top sheet was propolized to the top hive and the bottom sheet to the bottom hive, the two sheets weren't propolized to each other and slipped apart easily. Bees boiled out from between the sheets the instant I cracked the hives apart, but tilting the top one up revealed long openings between the frames. So I lowered it again hastily and squared it with the bottom hive, and tore off all the newspaper sticking out around the sides. Though I didn't notice till afterwards, there *were* crumbs of newspaper by the entrance. So they're through, and the two queens can settle (or already have settled) things between them, and I won't have to bother the bees again for a couple of weeks at least. Thank goodness.

A curious thing: right after I'd closed up the hive again, some robbing behavior commenced. It was too late in the day for Jim's bees to be over here; and since there weren't any more heaps of bodies on the ground, I think we can conclude that the *internal* union of the two colonies was peaceful. My hypothesis: the bees from the top hive are accepted by the house bees in the bottom hive if they go down through the inside; but if they fly around and try to go in through the entrance, they're perceived by the guard bees as robbers and attacked. The bees that spilled out of the top hive when I tilted it must have been the ones I saw being treated as robbers, when they were only trying to get back in.

It's time and then some to leave that hive alone. After I'd finished and come inside, Ted noticed a lot of bees on the patio, poking about among the covers and bottom board I'd left there, and in my box of equipment. More and more of them came. Altogether, this is the most agitated and aggressive I've known them to be, ever. Uniting the colonies was necessary to bring the weak hive through the winter, but the mortality from the whole procedure has been appalling. Dead bees are what the beekeeper wants most not to see, and I've seen (and smelled) thousands of them this week. I have to do better than this.

August 27. Muggy hot weather for the third day in a row. We decided to dig potato bed #1 this morning, partly because it was time, and partly to reclaim the topsoil from the pond excavation, to use for landscaping around the pond. The yield was disappointingly, but not unexpectedly, low; this is the bed where mice made so many tunnels under the roots of the plants. Bed #2 started off with less of a bang, but should prove more productive. Its plants are still green and growing, though, so we won't know for a while. I didn't spread the potatoes out on the patio to dry because Hurricane Andrew is expected to dump a lot of rain up here in another day or so. I may put them back in the spare room on newspapers; meantime they're still in a plastic dishpan, in which they're at least portable.

While we were digging potatoes Ted got his first sting of the season. The bee barrier has gotten saggier by degrees and some enterprising bee swooped through the saggiest place, bumped into Ted, and stung him on his bald scalp. Not a bad sting, luckily. All season he's managed to avoid this, till now. It might not have happened even now if the hive hadn't been so upset; but anyway I got the stepladder and tightened and retied the scrolled-up shade cloth.

Just as we were finishing, several more bees got interested in us, exhibited guard-bee behavior, buzzed around our heads, etc. Ted jogged off to the other side of the yard till he gave his the slip. I was so nearly done digging that I tried to finish, but the bee or bees harassed me till I decided it wasn't worth it. These may have been bees returning to the hive, accustomed to flying over the barrier at a lower level than was possible after I retied it, who were angry about being thwarted. Or, in their present riled-up state—they may have seen our activity from the hive and come over to object, though it's very unusual for them to go looking for trouble so far from home. Whatever the reason, I'm *very* glad the recent hive manipulations are concluded. Let the bees go about their business now and forget about me.

I changed the water in the duck tub so it would look nice for Edward and Bob, who were coming out this afternoon; but by the time they arrived it was opaque again and decorated with floating feathers. Oh well. Edward and Bob, friends who can't make the open house but were curious about the homestead, were *very* satisfactory visitors—intensely interested in and appreciative of everything they saw, asking lots of questions, really getting into feeding the fish, etc.

My five thousand redworms had arrived yesterday from Georgia, in a carton full of peat moss with holes punched in the sides; I added more water to the bedding and mixed it up some, but waited to introduce the worms to their new bin so Bob and Edward could witness the event. I must say the worms failed completely to live up to their name of red wigglers, and suspect the heat in transit may have been none too good for them. The box was mailed on August 25, meaning they made it here in two days, but they definitely didn't look all that lively. They lay in matted clumps of solid worms on top of the bedding. Mary Appelhof said at GardenFest! that they would wiggle down into the bedding and away from the light, but most of these worms did nothing but twitch faintly, if that.

But I covered them up and left them alone, first giving a little one to Ted to toss into the pond. ("Did they eat it?" "Did they ever!") Around 10:00 I took a flashlight and went out to see if the worms were still on top of the straw, in which case things would be dire indeed. But there were no worms in sight at all, so either something crawled in and ate them or they did what they were supposed to do and made their way down into the hog heaven of straw, soil, and duck poop prepared for them.

I took some duck-tub water over in the sprinkling can this afternoon and watered the two furrows of peas. While conducting the garden tour, around 6:00, I noticed that a few peas, invisible at noon, had poked through the surface.

The vines have pumped out still another basket of ripe tomatoes. A couple of the biggest were pressed onto Edward and Bob, along with four big cucumbers which I was especially eager to get rid of—we've already got enough pickles for the next year at least in the larder, most of them from just the two surviving Jazzer vines.

August 28. This morning the redworms were nowhere in sight. I stirred the bedding around until I found a couple of them, but I still don't know whether they got eaten, or if they just burrowed down and got to work eating stuff and getting bigger. The one torpid little worm I threw into the pond barely hit the surface before a bluegill snagged it. As I walked over, I saw a very small frog hop out of the water-hyacinth clump into the water, and heard another do the same with a hoarse little cry. How many have we got now?

The basket of tomatoes was giving off a powerful odor of rot. I sorted through the lot, washing off the good ones tainted with rot juice, cutting the rotten places out of others, filling a big bowl with pieces that went into the fridge.

A heavy, muggy day. Hard to summon up the energy for outdoor work. The remnants of Hurricane Andrew blew through in the evening, with high winds and rain, but our power stayed on so I guess we missed the worst of it.

Dinner tonight: eggs scrambled with corn, tomato and cuke salad. Exquisite! Now that the coleslaw season is over, I'll record for future reference the basic dressing recipe we never write down and then spend half the season trying to reinvent: fifty percent mayo to twenty-five percent sugar to twenty-five percent vinegar, plus mustard and celery seeds to taste, the whole thing well blended by mechanical device. What cabbage is left in the fridge the ducks will get. We're both pretty tired of coleslaw, and Ted's very tired of making it, and cleaning up the little cabbage fragments the food processor showers all over the kitchen.

August 29. Rain: 0.575 inches. After two dry weeks, not enough. In a few days, if we don't get more, I'll have to water.

Andrew blew down a few stalks of popcorn, loosened up the

just-tightened bee barrier, and overturned an empty compost can which it then dumped some water into, but the main thing it did was bring in a beautiful crisp blue day. For the first time in ages the bees aren't festooning the front of the hive. All appears normal and calm over there, finally, thank God. I've got at least six weeks in which to reduce three hive bodies and two shallow supers (one on the hive, one in the basement, both containing honey) to a two-story hive crammed with honey and bees—the best way to get them through the winter.

We definitely have a resident woodchuck. He putters innocently in the grass and is never seen to go near the garden. I regard the electric-fence wire with great satisfaction—probably he's already had his encounter with it. Bob asked, why wouldn't the woodchuck dig in under the gate? My answer: he might if he were able to figure out where it is, but woodchucks are no mental giants—except compared to, for instance, ducks.

The ducks, by the way, are very mopey. For a while this afternoon we put them in the garden, where they rested for several hours under the tomatoes. They ran from their pen toward the garden gate, flapping their wings like mad, evidently to dislodge the loose and itchy feathers. Then they went in and, after foraging halfheartedly for a few minutes, settled down under the tomatoes and stayed put. When allowed to return to the pen they went straight under their house. I'd thought it was at least partly the weather, but I guess it's the molt. The only thing they show true interest in these days is putzing around in the grass, and neither of us has time to stay out in the yard with them, to keep them away from the pond in one direction and the bees in the other.

One of the bluegills has a light discoloration of the scales on its head and back. Uh-oh. Lilypons may have to send me something to treat the water after all.

Ted has finished making some particularly thick and excellent spaghetti sauce, using two of our own small green peppers this time as well as a big one from the store. Fifteen pints, bringing the total to fifty-eight. A goodly number. About the Beltsville Experiment, by the way:

In the beginning I had some notion of counting the tomato totals from each bed and comparing them—good intentions that went down under the harvest tsunami. But judging by the way both beds have been pumping out tomatoes, vetch by itself is every bit as good a fertilizer for tomatoes as the fertilizer I used in the control bed. I don't know that it's better, but it's just as good.

The tomatoes in this last batch were exceptionally ripe and delicious, but if experience is any indicator the plants are about to poop out. The green fruits will stay green, and the harvest will sputter and stall. It happened like that the last two times; we'll see how it goes this year.

The redworms now look great compared to the shape they were in when they arrived—moist and lively in their bin full of duck bedding.

August 30. My fiftieth birthday, a spectacularly beautiful day. We marked the occasion by visiting the Brandywine Museum and Longwood Gardens—and by doing *no* work for me to summarize here this evening!

The Ideas Garden at Longwood—the only fruits and vegetables in the place—always provides a basis for comparison with my own garden. Today, as I was poking through the beds and muttering comments, Ted objected that I take it all so *personally*—that I needn't feel competitive with Longwood.

But taking it personally is what makes it *fun*. For instance, Longwood's cucumbers were foot-high seedlings. I knew what that meant—that the spring-planted ones had died of bacterial wilt—and felt smug about my own two Jazzers, still pumping out the cukes faster than we can consume them or give them away. Last year I wouldn't have recognized the bed of sweet potatoes; this time, of course, I knew instantly what they were. Longwood always seems to grow big healthy beets with no difficulty; mine invariably succumb to some foliage disease if the seedling roots don't get eaten by white grubs or the first leaves by slugs. Their dwarf fruit trees have the perfect shapes I try and consistently fail to achieve, and are covered with apples and peaches, suggesting that my trees must have been planted too deep to remember they're supposed to be dwarves.

Etc. Except for Liz, I'm without vegetable-gardening friends. At the Rodale and Longwood gardens I get to check out what other people are doing, and see whether they're doing it better or worse than I am. The verdict in each case: some worse, some better.

I recently came across a statement in *The "Have-More" Plan,* a 1940s book by Ed and Carolyn Robinson, from Garden Way Publishing. This is a belief I've never seen expressed elsewhere and agree with entirely. It's this: "Like many city people we thought a garden was 'duck soup'. But we've found out that our garden is our most exacting and complex project. Producing eggs, or chickens, or milk, or honey, or pork requires less knowledge than having a good garden . . . it's easier to produce a dozen eggs than a bunch of carrots."

Amen. This summer I've failed repeatedly to produce a bunch of carrots, but I expect to produce a dozen eggs next spring with very little extra effort. (I probably shouldn't say that out loud.)

September

September 1. After knocking off for the past couple of days—we spent yesterday, another gorgeous day, canoeing on the Brandywine—I got up this morning and made a long, serious list of Things To Do:

duck muck and siphon
pick
    tomatoes
    cukes
    corn
    apples
cut secondary cabbage heads to two per plant
pull up any cabbages with rot (don't compost)
fix collapsed popcorn
cut broken sweet-corn stalks
fix fence post
water peas with duck water
read up on and medicate fish
read up on popcorn harvest—leave to dry on stalk?
weed blueberries
call Lilypons about
    fish
    cattail
    Comanche
get b&w film
check worms
SPRAY BEANS!
GET MUMS!
weed electric fence
set up new cold frame

    I tore through this list like a house afire, all the way down to SPRAY BEANS!—which I clearly want so firmly not to do that I managed to forget again, despite the capitals and exclamation point—and

GET MUMS! which I decided against after finding out how much it would cost to fill the rest of the former onion bed with them. Both weeding the fence line and setting up the cold frame are bigger jobs than this day allowed space for; but everything else actually got done, or as much done as was possible.

As expected, the tomatoes are slowing down. The sweet corn is finished. The cukes keep tumbling in; we have enormous cucumber-onion-and-tomato salads every night, but neither of us feels much like canning any more pickles.

When I removed the lids from the worm bins and raked through the bedding, I saw only one garden worm and maybe half a dozen little redworms, total. Something (a mouse? a shrew?) had dug into the garden-worm bin under the side. The garbage I tucked into the redworm bin yesterday had not attracted any takers. They're (1) dying, (2) escaping, or (3) being eaten, but what they're not doing is thriving in the duck straw. Looks like I'll have to order more worms and set up a conventional box indoors. This outdoor worm farm would appear to be one speedy failure, and for reasons which are—and seem likely to remain—a mystery. If I try the experiment again next year, it had better be more carefully thought through.

At Liz's suggestion I called the hatchery where I bought the bluegills and catfish, and described my fish malady, but they were unhelpful. The Lilypons catalog lists a lot of goldfish diseases but nothing that sounded much like this one; so I called them—I'd been meaning to anyway—and did receive a helpful suggestion: try rock salt. A weird idea, to salinify the pond water. Something to do with electrolytes, apparently.

The application is a pound of salt to every hundred gallons of water, at which rate I'd have to put in a hell of a lot of salt—but it's a better idea than medicating the entire pond, and I'd already spent about forty-five minutes trying in vain to "net" the sick fish with a sieve. Liz had brought over a bottle of DesaFin, a remedy for every fish ailment under the sun, and my idea had been to isolate my problem fish in a bucket and medicate the bucket; but I couldn't *catch* the aggravating thing. By holding the sieve underwater and dropping shrimp pellets into it I actually got the right fish into the sieve one time, along with one other one, but before I could get them into the bucket, one flipped out. Guess which one. And I never again managed to catch it. The blotchy fish is timid and hangs back from the feeding crowd. Again and again I had almost lured it in when a bossy bigger fish chased it away.

As my hopes of catching it faded, I started to consider medicating the whole pond. But then Ted read on the box that DesaFin should not be used on fish intended for human consumption. In that context, having another course to try looks real good. So, as soon as I can, I'll figure out my volume and go buy a bag of rock salt adequate to the application.

I reported the deaths of the Comanche and the cattail to Lilypons, who promised to replace them in the spring.

Despite my deep conviction that the homestead has *already* been a success no future calamity can neutralize, I still feel terrible when one of the experiments fails, and two going bad on the same day is pretty hard to take. But the point bears repeating: whatever happens now, the homestead will have been a success. Nothing can change that basic fact, though how qualified a success is still to be determined by the events of the next couple of months.

Certain of the experiments have flopped: the screenhouse produced no melons, the worm bins were a total bust as secure containers for worm breeding, the greenhouse never got built, the fish won't be big enough to eat this year, certain crops (garlic, carrots) and flowers (sweet peas, marigolds) never got to first base. But the freezer is filling up with vegetables and spaghetti sauce, there are still most of fifty pounds of honey in the basement, plus a year's supply of pickles and blueberry jam, and three more ducks will appear on the table before we're through.

Most of all, the whole system works in an integrated way. With some refinements, which are easy to imagine and wouldn't be hard to implement, this year's kinks would come out.

But the truth is, even if the Utah plan hadn't developed, it doesn't feel sensible to me to go on trying to do this in suburbia. It's too unnatural; there's a fundamental absurdity to the idea of subsistence living and economic independence in a posh suburban setting. As a one- or two-year experiment, aimed at acquiring skills and learning how things work, it's a great idea; but *here* isn't where living this way works economically. The land and houses are too expensive, the values are too concerned with cosmetics. Country life should be lived in the country.

I'd like to have carried the experiment forward one more year, but then to dismantle the homestead here, maybe to reconstruct it someday in a more appropriate place (not Utah) . . . Learning what is and isn't appropriate is also part of what the year's been about.

September 2. Funny how its disfigurement makes that one fish suddenly an individual, distinguishable from the school of bluegills for the first time. I tried again to catch it, an exercise in frustration. Tying the sieve handle to the rake handle gave me more reach but also more weight and less maneuverability. What's needed is a shorter handle than the rake's.

I gave up watching the fish's splotched little head retreat from the bullying bluegill's and turned to more possible projects. Inspired by an article in *Country Journal* about cold frames, I'd decided to reconfigure Boxed Bed #1 as a four-by-eight-foot cold frame. Accordingly, I dragged the boards out of the jungle and selected the four in the best shape. After we dug the potatoes the other day, Ted carted the topsoil back to the pond area, so the site was bare earth—much of it clay, but dry and friable, easy to turn. I spread most of a garbage can's worth of compost over the area and dug it in, then set the four boards in place and laid two aluminum-framed glass louvers over the box. These louvers, the same ones I used as the top of the solar wax melter, are four feet long by one foot tall, a perfect fit. If there are six more unbroken ones in the garage, we're in business.

Raked smooth, the soil inside the box comes so high up the boards that I realized I should put the four *worst* rotted boards on the bottom and perch these on top, to give the growing crops some head-room. But by then it was time to get dressed and catch the train into town for my final birthday remembrance: a trip by ferry over the Delaware to the New Jersey Aquarium.

Bees have increasingly been mobbing the water clover. In the middle of these sunny days I'm reluctant to go back by the pond. They're interested in the water, not me, but with so many buzzing around, I'd just as soon stay out of their way. Some of these bees have black abdomen tips, some are striped. The black ones may be Jim's, but in any case aren't mine; mine are all classic striped Italians. Don't know if we still have five frogs; only two jump in when we approach these days.

Note: the ducks are extremely molty and miserable. They stay under their house all day and emerge looking more rumpled every time. Not even bread tempts them these days. They never clean up their feed. Poor things, I hope this doesn't last much longer. The whole pen is full of loose feathers, and the tub is filthy again. I still can't tell who's a drake and who's a duck, but it seems to me that only one of these birds is quacking. If it turns out that only two of the seven were females and we ate one of those, my plans for keeping two pairs over the winter will be knocked into a cocked hat.

September 4. A collector in California has sent a box of my novels, along with several magazines containing stories of mine, for me to sign. It was a strange half-hour I spent this morning, unpacking, inscribing, and repacking all this material. You have to be nice to the fans, and anyway it was pleasant to think of myself as a collectible author; but the dis-junction between that short interval and what I otherwise do with my time these days felt odd. Even *Time, Like an Ever-Rolling Stream* turned up in the box (already!). The hardcover *Ragged World* was in mint con-dition, obviously never even opened, let alone read. What makes col-lectors do what they do, I wonder? Book As Object isn't a concept most writers can relate to very happily. Still, this guy had *bought* the objects, so who's complaining?

Later, desk cleared, it was back to farming.

The "experiment" of letting tomatoes sprawl on the ground vs. caging them has come out overwhelmingly in favor of cages. Nearly every tomato rots if it actually touches the soil. Slugs break the skin, bacteria get in, in no time what was a ripe tomato is transformed into a sphere of smelly mush.

The nightshade report: (1) I picked more tomatoes today. Either Ted will have to make sauce tomorrow, or the ones picked Tuesday will have to go in the fridge and suffer loss of flavor. (2) Three green peppers were big enough to bring in with the tomatoes; but, as usual, the main pepper crop will completely miss the main tomato crop. (3) Tomorrow without fail I'll have to make ratatouille, or our two eggplants will be too old and soft to use.

This afternoon I marched myself out with my sprayer of Bt and doused the brussels sprouts; and at dusk, after *eight days* of procrastinating and forgetting and being forestalled by weather, I finally sprayed the beans. The wax beans have made a nice comeback from the earlier bean-beetle assault; their poles are covered with new, unchewed foliage and there are plenty of flowers. I'm not sure why I dislike spraying so much, but one reason I kept not doing it is that it has to be done at dusk, "to spare [as they always put it] the bees." Forget to spray at dusk and you have to wait twenty-four hours for another chance. Rotenone acts and breaks down quickly. By the time the bees are active tomorrow, the bean beetles should all be dead, adults and larvae alike, *and* the bees should be safe.

Nothing new to report about the livestock. The ducks are still molty, the bees still uncomfortably thick on the water clover, the worms more disappeared than ever—today I could find only one redworm in the bin. I did see two frogs, not that they count as livestock exactly. The blotchy fish is still blotchy. I called Lilypons again and discussed the subject of rock salt in a little more depth. Apparently the rock salt they mean is the kind used to make homemade ice cream, isn't sodium chloride, and won't salinify the water and kill all the plants.

The pond water is the color of weak tea, but very clear. Pictures that came back recently show the water as absolutely colorless not so long ago. Tannin dissolved from the sunken leaves may have tinted it like this.

Maybe the other fish won't get blotchy; or maybe the hatchery people were right, and the discoloration resulted from an injury.

Only 0.2 inches of rain in the gauge this morning, but the pond was nicely topped up. I wonder if the gauge is giving me the straight story. Too many pine trees too nearby, I shouldn't wonder.

September 5. I overslept this morning and staggered into the kitchen shortly before 10:00 to find a caldron of red sauce already boiling away on the stove. By noon this batch was finished: 11 pints, for a grand total of 69. Probably this was the last batch; we don't need any more and neither of us will have time to make more, with the semester kicking off next week.

And with us off to Utah again in a few days, to buy a house. I'm going out first to narrow the field; then Ted will join me for the final decision, if any. The housing market is so hot out there that prices can only get higher the longer we wait. So we're going to try to find something suitable now, install a tenant for the duration, and move in a couple of years.

I delivered open-house invitations to the neighbors today, and a big Jazzer cucumber with each invitation. The cukes were greeted with cries of dismay: "Are those *zucchinis?*" Liam agreed to go back to work for four days while we're away, and came over to be apprised of the minor changes in the setup (bucket instead of waterer, two pens instead of just one). I'll write him up another page of instructions and confidently leave the cranky ducks in his custody.

Stacking the two boxed-bed boxes on top of each other to make a deeper cold frame isn't going to work. The timbers already framing the bed are the best of the lot, but are pretty rotten on the bottom themselves. I raked the bed as smooth as I could and planted all the remaining Narova and Scarlet Nantes carrot seed, and more Black-Seeded Simpson lettuce. Then I poured a sprinkling can full of duck water into the bed and dusted diatomaceous earth in the rows prior to putting the glass louvers on top. Naturally the d.e. absorbed all the water I'd just put in and was rendered instantly ineffectual against slugs. Dumb. Let the soil dry out overnight at the surface, then try again.

Can't cover the top entirely with the weather still this warm or I'll get a solarized bed instead of a better microclimate, so the louvers are spaced like gaps in teeth: air in, most rain out. We'd have sent these louvers off to the dump if Bill hadn't suggested they'd be easier for me to handle than the lead-heavy double-glazed patio doors. So, thanks to him, I have eight one-foot by four-foot panes of glass framed in aluminum, which cover the box perfectly and are a snap to move around. Bill understands salvage.

When I get back, in a week, the seedlings should all be up and ready to be dusted again with diatomaceous earth. Then I'll have to monitor vigorously, cover the box to keep the rain out and the diatomaceous earth dry and efficacious. *And* water everything with duck water, till the bed's too well established for slugs to make any difference.

We're going to have to cover the pond with netting—two layers, to make meshes small enough to keep out the walnut leaves. I glanced up this morning just as a breeze blew what looked like several bushels of leaves off one of the trees. It won't do; we need to try to prevent all that walnut stain from getting into the water, and all that biomass from winding up eventually on the bottom.

The garbage I put in the redworm bin the other day has attracted gadzillions of millipedes but *no* redworms. I raked all through the bedding without turning up a single one.

The bees have become a genuine menace at the pond. They hang in massive clusters on the front of the hive, with nothing to do in the absence of a honeyflow—the same reason they hang out at the pond all the time, I'm sure. They'll probably go on like this till the weather turns chilly. Shouldn't be long now, but we could have a bee problem at the open house on the 19th.

The surviving Comanche water lily bloomed today, a beautiful creamy yellow flower like those we saw in the Longwood water-lily gardens a week ago. A second bud has nearly broken the surface. The other Comanche is totally dead; I hauled its container out of the water so it wouldn't decompose in there and dumped the contents into the bushes, clay, gravel, tuber and all.

September 6. An all-day drizzle today meant I didn't get far with my list of pre-travel good intentions. Only 0.2 inch in the gauge by late afternoon, but this was no day for putting netting over the pond or setting up pea fencing, and Ted sought in vain for rock salt at the supermarkets yesterday. Blotchy looks no worse and doesn't act sick, less timid if anything than formerly. Unless the condition deteriorates by the time I get back, or other bluegills show signs of developing the same condition, I guess I'll go with the injury theory and do nothing. (Ted says: Maybe one of the catfish attacked Blotchy, springing out of the undergrowth, snarling, claws bared . . .)

I put the rumpled ducks in the garden and did the duckmuck and siphon, tramping around in the drizzle in my black Wellies, with cart and rake and duckmuck bucket and armful of clean straw. I'd hoped that by putting it off till the last minute, things wouldn't get too awful in there by a week from today. They've been spending most of their time under the house, which concentrates the smell wonderfully. They've also been playing in the mud under the downspout, which guarantees that the clean water in the tub will be black by lunchtime tomorrow.

I poured some of the muckiest tub water into the hyacinth holding tank, where the sidelined hyacinths have been looking very chlorotic and puny; the rest went on the peas. Chlorotic hyacinths put back into the pond don't green back up, but the new foliage they put out is vivid green. Liz thinks the bees are attracted to the yellowed leaves out of desperation about their flowerless state, and took me to task last week for not growing something that would be in bloom this far into the season, to tide them over. But they've got scads of stored honey, they're okay. Even in the rain, though, they were drinking from the mat of water clover.

I'm concerned about leaving the pond netless, especially after Ted comes out to join me in Salt Lake, but that's the way it is. If the water gets too problematically dark and leafy, we can always (groan) pump it out and clean things up, later in the fall.

For dinner tonight: ratatouille! I was determined not to waste those eggplants, but then put off making it and put off making it till finally Ted started to help and then virtually took over. We used a recipe for classic ratatouille from *Stocking Up III,* leaving out the zucchini and increasing the amount of eggplant. Somewhat to our surprise it was perfectly presentable ratatouille, made all of our own ingredients (eggplants, tomatoes, peppers, onions) except the herbs, and served with some of our heated-up leftover potatoes on the side. Mmm. Now we know what to do with all the eggplants that seem about to inundate us. N.B.: there's no point whatever in growing these as container plants; the in-ground pair are huge and healthy—much bigger than the two in containers—and covered with flowers and little fruits. There must be *one* soil-borne disease we haven't got!

September 15. I had to be in town all day yesterday and today, and therefore have done almost nothing about the homestead since our return from Utah on the 13th. (Where we did find and buy a house.) And lots of things need doing. The open house, only four days away, threatens to be a disaster. The duck pen reeks from nine days' accumulated guck, the algae on the pond liner is an inch thick with no scavenging snails' trails to be seen on it, the garden is a pale shadow of its former glory, with many rotting tomatoes on the ground, there's a stench from the hive that I'm scared might be foulbrood, the electric fence and the potato beds desperately need to be de-thistled . . .

Well, we did get started. Ted, who was home today, put in four hours' work cutting the grass: instant morale boost. After I got home from teaching my first class, we did the part of the mowing that takes two people: the solar panel and battery, the garden paths, the asparagus. I move the vines and hoses out of the way while Ted mows the garden paths; then he holds the asparagus up with a duck-herding wand, while I mow in my bee suit. Then I cut the rest of the grass around the hive(s).

As soon as I got over near the hive, the mower threw up the dismayingly familiar stench of rot. The source could be the new heaps I found when I mowed before the trip, ten days ago. I better hope so. I also better look in tomorrow to see what's what. It wouldn't do to encounter an unpleasant surprise on Saturday, when I open the hive for an audience.

After dinner tonight (spaghetti and salad) I cut up the tomatoes Ted picked before joining me in Utah, plus the ones I picked yesterday. The last two years we grew tomatoes they had quit by now, but not this year; the cut-up pieces completely filled the big pot. The plants look godawful, diseased and bald below, but they keep putting out ripe tomatoes anyway, somehow. One of the Jazzer cukes is also dying—not of wilt, but of some disease that yellows the foliage, probably mosaic—and so is one of the in-ground eggplants. More potatoes are dying too, but that's normal. Things look extremely ratty though, for company to see. It's too late by two or three weeks to try to show off a garden. Too bad I never planted any pumpkins.

The soil in the cold frame is cracked, and the diatomaceous earth looks poisonous, but lettuce and carrots both came up pretty well in there. They badly need water, which they'll get tomorrow—I didn't want to water them at night because of the danger from slugs, and didn't have time either today or yesterday before dashing off to catch the train. The peas are tattered but hanging in there; they need a fence more than they need water. The pepper plants are finally producing nice big bell peppers, and the brussels sprouts are bigger, too. So there'll be a few things to look at, but mostly it's a chewed and motley garden people will be traveling from as far away as New York to see on Saturday (if we aren't rained out).

The pond. I looked in vain for the blotchy fish; either it died or it got better. The algae bloom is probably the result of the fact that the fish are getting bigger and excreting more nitrogen, which my reduced number of water hyacinths can no longer take up sufficiently. All the submerged plants are fuzzed with algae too. I tossed a few of the anemic water hyacinths from the tub back into the pond, noting that the ones put in before I left have started growing nicely again.

The leaf problem seems less severe than the algae problem just at present. And some of the bluegills look positively huge, twice the size of others; can they have grown that remarkably in nine days, on a diet of mostly algae? Ted saw *two* catfish today, going after the chow that sank to the bottom and chasing the bluegills away. The cats started out bigger, and they're still bigger. Whether we still have tadpoles or frogs I do not know; it'll take better observation than I've been able to do to ascertain that. Maybe tomorrow, when I plan to spend the whole day working out there.

I made fifty copies of a Self-Guided Tour Map of the homestead on the English Department copier. (This is criminal behavior.) I need now to draw an enlargement of just the garden and make fifty of that, and staple them together.

Things are winding up. I feel it very strongly, even though my time in the yard has been so brief. They're winding up, or down. I'm glad, I guess, but it's sad, too.

Liam's mother, Lynn Castellan, called Sunday, as soon as she'd noticed we were back. Ted talked to her and reported afterwards, "We did a terrible thing. We forgot to tell Liam there were only six ducks." The poor kid had worried the whole time we were away that *he* had somehow caused the seventh duck to disappear. He came over a bit later and said—very likely quoting his father—that he'd thought maybe it was at the bottom of the bathtub. It never occurred to us that he wouldn't know the seventh duck had entered the Slug-Hunting Ground in the Sky a long time ago, but we never actually informed him of it; the neighborhood kids all have trouble with the idea of eating the ducks, and I guess neither Ted nor I was eager to broadcast the news. But we hadn't realized that none of the Castellans knew.

Before penning up the ducks tonight, I stuck the two hoses behind the fence and started the siphon going. The tub's the source of at least half the stink out there—dark brown water with lots of black feathers floating in it. Ugh. Early tomorrow I'll muck out the house; that should improve things a lot. It'll have to be done again Saturday morning, but it certainly can't wait till then.

September 16. We have at least two frogs, both little. No sign of any tadpoles.

This was a day of heavy, productive labor; I weeded bushels of thistles out of the potatoes and more from around the electric fence line, tied up the sagging bee barrier more securely, and pruned and tied up the kiwis—a long-overdue job.

I also gave an interview. A reporter named Reid Kanaley from the *Inquirer* called to ask if he could come over. I worked right up to the minute he arrived, and worked some more after he left, but talking to him about the homestead was also work. He was here for an hour and a half, from 1:30 till 3:00, and afterwards I was too bushed to continue till I'd had my tea.

Doing the interview was good for my morale. A while back I was very keen to get the homestead into the news; now things have moved ahead and I no longer felt all that interested in the idea, in prospect. But showing it all to Reid, and explaining how the various parts are interconnected, got my juices flowing again. Now I think the open house will be fine, if only it doesn't rain. Things don't look as good as they did, but you can still tell how terrific the place and the project are.

I rooted through the worm bins. One garden worm, infinite number of slugs, millipedes, and sow bugs gathered where the garbage was, but not one single redworm of the five thousand installed in the redworm bin. Without having any way at all of knowing what really happened, I'm taking the official line that they escaped. The bin idea may be sound, but the next bins will have to have bottoms.

Despite the arduous day, I did take a little time out to sit on the bench and observe the pond, with the result that I've made a command decision. Until further notice, nobody is to feed the fish. They've grown so well on shrimp pellets and bread that they're putting out too much nitrogen for the plants to take up, which is obviously what's caused the algae bloom. If we stop feeding, the fish'll have to live on algae for a while, and the problem should correct itself.

I see no evidence of any surviving snails, for no reason I can imagine—certainly not lack of algae!

September 17. The *Inquirer* sent a photographer, named Beverly, over to take pictures to illustrate the article, which will be, not in the *Neighbors* section, but in *Suburban/Metro*—hence not just local, but regional. She took duck-herding and -foraging pictures, weeding pictures, white- and sweet-potato-digging pictures, fish-catching pictures, bean-picking pictures, cart-pulling pictures, and finally beehive-working pictures. For an hour and a half she snapped away, and by the time she rushed off to develop her film I was just as pooped as I was yesterday.

Really exciting to unearth the very first sweet potatoes, looking big and beautiful. Not so great "netting" a bluegill with the sieve. I was able to catch a couple, but they flopped about so violently in the metal mesh that I don't think I'd better try that again. I should get a proper fishnet that they can't hurt themselves on. Blotchy has turned up, by the way, looking perfectly healthy, and with her blotches perhaps a bit faded.

I weeded the pond area and raked the heaps of topsoil around, then distributed the compost evenly and raked that around too. We can put boards or something down for people to walk on on Saturday, and seed the area with grass seed on Sunday. It doesn't look great, but it does look somewhat better—more purposeful, if not finished.

Ted's not feeling well and went to bed as soon as he got home from town. I hope it's only exhaustion and not a bug, both for his sake and for the party's.

September 21. The open house has totally dominated the last few days, so this will be a collective report, starting with Friday the 18th, the day the article appeared in the paper. Not just in the suburban edition's *Metro* section, either. They also ran a somewhat shorter version in the regular *Metro* section, so millions of people may have seen the photo of the six silhouetted ducks filing toward the garden, with the small figure of the Goosegirl in the background. A wonderful picture, and a very good article, marred only slightly by Reid's saying that I feel we've "trashed a stack of *snooty* social contracts" in pursuit of our home-steading goals, after I'd told him the neighbors had been great. My tone was rueful rather than nose-thumbing, but oh well, I'm still happy about the way the story turned out.

Apart from freezing a few beans I'd picked the evening before, the whole day Friday was devoted to planning and shopping for the open house (and to turning on The Weather Channel nervously to check on the most recent forecast). The outlook improved as the day wore along, and finally it began to look like a pretty good bet for Sat-urday. And in the event, Saturday was a splendid day. It started off cold and dark, but by noon the sky was brighter, and by 2:00 the sun was out and everything looked beautiful.

I'm still amazed to think it, but by all accounts the party was a terrific success. About forty people came, I believe, though I didn't think to keep a guest register till too late. They were, as Ted observed, a mot-ley crew, consisting as they did of the Philadelphia science-fiction crowd, some Penn-associated people, four families of neighbors, and a sort of "miscellaneous" group made up of my father and stepmother, Sheila Williams from *Asimov's*, Bill Wasch and his family, Liz Ball, and probably some others I can't think of right now. Dad and Betty drove all the way from Simpsonville, Kentucky, for the occasion. Cornelia Hasel-berger remarked later that it was the most *unusual* collection of people she'd seen since coming to this country from Germany a year ago.

The remarkable thing was how well everybody seemed to mix and get along, all without any alcohol to lubricate the occasion. People wandered about reading the signs Dad and I had put up in strategic lo-cations around the garden and yard—based on questions Ted and I get asked over and over again—and consulting the three-page handouts I'd made showing (1) the plat of the yard, (2) a blowup of the garden with growing beds labeled, and (3) How It Works: the interrelatedness of all

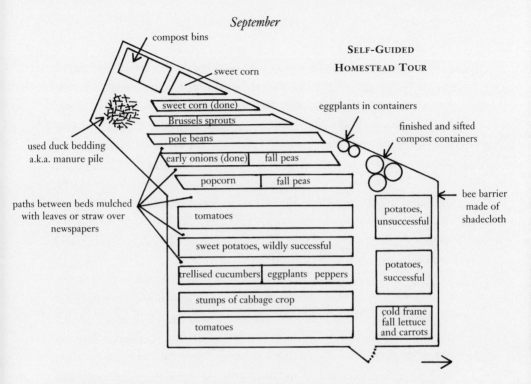

compost bins

SELF-GUIDED
HOMESTEAD TOUR

sweet corn

sweet corn (done)

Brussels sprouts

pole beans

early onions (done) · fall peas

popcorn · fall peas

tomatoes

sweet potatoes, wildly successful

trellised cucumbers · eggplants · peppers

stumps of cabbage crop

tomatoes

eggplants in containers

finished and sifted
compost containers

used duck bedding
a.k.a. manure pile

paths between beds mulched
with leaves or straw over
newspapers

potatoes,
unsuccessful

potatoes,
successful

cold frame
fall lettuce
and carrots

bee barrier
made of
shadecloth

the parts. As they wandered, they bumped into each other and introduced themselves. I'd provided name tags, and people thought of funny
things to put on them, with Betty's help, especially the science-fiction
gang: Gardner Dozois was "Organic Gardner," Greg Frost was "Walking Compost," Ted was "Migrant Worker," etc. Bill's was suggested by
Liz: "Pergola Builder" (she liked the job he did on ours and has hired
him to build one for her own garden).

Vicki and Liz managed the food, to perfection, bringing out
platters of doughnuts and jugs of cider whenever more were needed.
Really, between them, with Betty's help, they did most of my hostess-
ing job for me. Vicki had brought not only the twelve dozen doughnuts
of assorted kinds, but sandwiches for Ted, Dad, Betty, me, and herself,
which were so generous they provided dinner as well as lunch. Ted *did*
have a tummy bug and couldn't help with the food, but the sandwich
went down anyway. He kept crashing, then rallying, then crashing; but
he loves a party and managed to get adrenalized enough to be sociable
for the second half of this one. (Then he *really* crashed.)

The Sun Chips and strawberries vanished in no time. I hadn't provided enough cups—the kids were tearing around, sword-fighting and climbing on the woodpile back in the jungle, and got sweaty and thirsty and would get a new cup every time they came back to tank up again. But all that side of things was taken out of my hands, so I managed not to get in a swivet about it, once 2:00 had come.

Ted had thumbtacked a copy of the *Inquirer* article to a board, and hung the board on the pin oak by a spike, so people who hadn't seen the article in the paper could read it. Everybody seemed to assume the timing of the piece had been deliberate—as if I had any control over such a thing!—but it *was* a happy coincidence.

I opened the beehive at 3:00, which as luck would have it was exactly when about half the guests arrived in a bunch. In the pressure of the moment, with several dozen people watching, I did something I'd never done before: I let a frame slip out of the frame gripper. Embarrassing as this was, it provoked a lot of questions about bees, and that saved me from feeling too hopelessly ridiculous. Later Ted and I herded the ducks into the garden, but I put them back in their pen almost at once because the little boys were fascinated, especially Ben Purbrick and Sean Swanwick, and kept wanting to get close and pet them. The poor ducks, of course, thought they were being stalked and got hysterical. So back they went, and immediately fled under their house and stayed there, till near the end when they seemed to get used to the commotion and came out for a swim.

Liz shot two rolls of film—72 exposures—so I should have a great record of the day. Ted took some pictures too.

People started leaving around 5:00, but almost all of them, no matter when they'd arrived, stayed till the party was over. I went around the whole time in a kind of fragmented daze, wanting to talk to everybody and getting only a few minutes if that with each person, and the activity kind of swept over and around me. But I have a lasting impression of brightness and chatter; and I don't know how many people told me how much they liked other people they met, which seems the best evidence that the party was indeed a big success.

Afterwards I was done in. I'd expected that Dad and Betty would have dinner with us, but they'd put in a long day too, and with Ted out of action and me so wiped out, we decided to forgo dinner.

Yesterday I slept in, then spent most of the day cleaning up, returning the chairs borrowed from the Stipes, taking down the signs, that sort of thing. I also spent it trying to process the experience of having given a large, successful party focusing on a subject of educational and philosophical importance, which I'd devoted a year of my life to exploring. The open house was clearly a Very Big Deal, a marker event requiring some time to get used to, but today I seem to be getting back to normal—not a moment too soon, either; tomorrow I have to teach.

Today I went around and picked everything I'd been leaving in the garden just till the party was over: cukes, tomatoes, peppers. Some eggplants are ready to come in too, but I didn't have a knife, and haven't bothered to go back for them. A couple more days won't hurt, and I *had* to get my class ready for tomorrow. And that's where the rest of this day has gone.

I opened the hive to have a thorough look at things, try to determine if the bees were acting demoralized (which they would be if, for instance, the queen had been killed Saturday when I dropped the frame), and put the super of medicated honey, with the cappings scratched, above an inner cover, so the bees would move the honey down into the hive and I could take the super off. As far as I can tell, things are proceeding normally in there. Numerous frames of capped brood in the bottom box along with some healthy-looking uncapped larvae, and a fair amount of uncapped honey in the top box.

On Saturday something ended, with a bright clash of cymbals and the dousing of a flame, three days before the autumn equinox and six months from the time I started keeping this detailed record. Due public notice was taken that something important had happened and had been brought to a fitting conclusion. It hasn't really *concluded* yet, of course, but I doubt there'll be much point in keeping a daily journal from now on, into the fall. The year has turned; the results are in. The experiment is successful—for me and Ted, for the friends who've followed its progress all this year, for the forty-three people who connected with it on Saturday, and maybe even a little bit for readers of the *Inquirer* article who will never meet me or see the homestead. *You have to let them come*, Harlan said, and he was right. Don't ask me how I know. I just do.

September 24. The weather has changed decisively and deliciously into autumn: very cool, breezy, hard blue sky, rosy golden sunset. It's wonderful weather for outdoor work, or would be, but I'm so TIRED! I pushed through the party and cleanup and Tuesday's class and errands, and then sort of collapsed. I feed the ducks, pick tomatoes, pay bills—minimal chores—but keep lapsing into stupors and reveries I have a lot of trouble snapping out of. *Tired.*

Still, other things do get done. The filthy duck tub gets siphoned and refilled, the popcorn—25 stunningly beautiful ears, all different, mostly deep russet or a mix of purple, white, and yellow kernels round and shiny as gemstones—gets harvested and laid to dry on the patio with their husks peeled back. The basement, after how many months of talking about it, gets cleaned. The hive gets checked. (Whoever wrote that bees will carry honey from a super placed above an inner cover down into the hive, if you scratch the cappings, was mistaken; my bees just repaired the scratches.)

How do I condense a three-story hive, with one super on and another super waiting in the basement, into a two-story hive for winter? I don't know, and don't seem able to summon the energy to work on the problem. I'm *really* tired. I can't imagine—for instance—how I'll find the wherewithal to slaughter and skin a couple of ducks, though that must be done soon, or even to rig a net over the pond, another pressing task. The walnut trees are nearly bare now, but the oak leaves are starting to fall.

The pond water has cleared up amazingly in the past two weeks, undoubtedly thanks to the ratty but hungry water hyacinths, and the ducks are increasingly beautiful as their adult plumage comes in and the green-and-purple iridescence becomes more and more pronounced. At least two have the telltale curly tail feather that means they're males. They've cheered up quite a bit as well. Today I put them in the garden while cleaning their tub; when I went to fetch them, I found they'd pushed over one of their barriers and were back in the corn patch by the compost bin. The first time in a month at least that they've shown any spunk. All through the molt they hung out sulkily by the gate, quacking to go back to their pen, so they could crawl under the house and be miserable.

I sympathize. I'm not miserable, far from it, but I wish I could (so to speak) hide out under the house for a few days, catch my breath, get rested and calm enough to pick up the ball of this experiment and trot on again.

October

October 7. Immediately after the above came a week during which farming gave way almost entirely to traveling (Washington, New York), teaching, grant-application completing, and general desk cleanup (still ongoing but near its end). The diverse and multiple claims of my other lives, honored at last. Following this, on the first of October, I walked around the yard with a clipboard and made a long, long list of Things To Do. Then over the past week I worked my way doggedly through the list. Here it is:

> duckmuck and siphon
> pick and clean dry beans
> pick tomatoes, peppers, eggplants,
> make and freeze ratatouille
> dig potatoes
> dig sweet potatoes
> set brush for peas to grow on
> spray brussels sprouts with Bt
> put away screenhouse
> clean dead leaves and lily pads out of pond
> rig netting over pond, read up on winterizing
> order menthol for bees
> reduce hive to two boxes
> feed Fumadil and Terramycin

It's taken me most of the week, but nearly all these harvest-season chores are done. When I disassembled the parts of the screenhouse, which had partly fallen down anyway in a big windstorm, what did I discover under the collapsed pieces of screening but a tiny melon, started at last! Shall I pot the plant and bring it inside, or build a cold frame around it? Would either strategy allow the melon to mature? I have to try. Weighing heat against light, I guess I opt for heat: potting the plant up, bringing it inside. Maybe tomorrow.

Tomorrow, though, a complicated beehive manipulation.

When slightly scratched cappings failed to induce the bees to take the honey down into the hive, I went to the other extreme. I slipped the bee escape into the cover below the super, which cleared the super of bees pretty effectively, then brought the super into the kitchen (along with maybe a dozen bees) and uncapped all the frames of honey. I don't know why I thought this could be done so quickly and neatly that it wouldn't be necessary to put down any newspapers; it made an unholy mess of honey and wax on the kitchen and dining-room floors. I had to scrub them on my hands and knees. But once the super of uncapped frames was back on the hive, they cleaned it out in no time. So I brought that one in, uncapped the one from the basement, and put *it* back on the hives.

While waiting in the basement, a pool of honey had drained onto the newspapers beneath this second super. I weighed the risks of initiating robbing, but couldn't bear to let so much good honey go to waste and took the papers out to the hive. At dusk I went to pick them up. The newspapers were perfectly dry, though stained where honey had soaked in—and on them lay a very small dead mouse. He'd been robbing the honey and suffered the same fate as that of robber bees; he'd been stung to death. Two dead bees were still attached to the stingers in his body.

This afternoon I slipped the bee escape underneath this super; by tomorrow it should be clean and cleared of bees, and I'll take it off too.

At that point the hive will consist of three hive bodies and be a lot better furnished with honey than before. Then the task becomes simply to put the nine frames containing the most brood into the bottommost box, put the rest of the brood into the middle box, and fill that box out with capped honey. Then the inner cover goes on. Uncapped honey can go in the third, and topmost, box, and be set above the inner cover, to be taken down below. And once all the frames in that third hive body are cleaned out, I can remove them from the box and invert a feeder jar of medicated syrup inside it, over the hole in the inner cover.

Fumadil is for nosema, Terramycin for foulbrood. Menthol is for tracheal mites. For no reason I can figure out, I seem to have a terrible resistance to ordering any menthol. Jim, who gave me an extra bag of it last year, tells me menthol should have been put on the hive in early September, so the fumes would spread while the weather was still warm. Probably all the hives in this part of the country are more at risk of winterkill from mites than from foulbrood, and with all the robbing to transmit pests between hives it's crazy not to order some menthol—and yet I don't.

The rest of the potato harvest weighed in at around sixteen pounds, bringing the season total to twenty-five pounds more or less. The most productive variety was a white potato called Ontario. By a coincidence, the bed of sweet potatoes also came to sixteen pounds or thereabouts. They were a pleasant crop to raise—almost no work beyond a little watering, and a satisfactory showing to justify their space. They pretty much fill a plastic crate left over from the pond planting. You're supposed to put them in a container, with all the adhering dirt still on, and then not disturb them again, except I guess to take a few off the top from time to time and eat them. The variety called Jewell is a strong purple-red, a color I never saw in a sweet potato before. This week they're being sweetened by living upstairs in a warm spot by the patio doors; next week they'll go down in the basement for the winter. The root cellar I always wanted would come in handy now, but sweet potatoes ought to keep well enough even so.

A spate of rainy weather, coinciding with my being out of town, did in the cold-frame lettuce and carrots. Slugs in their multitudes must have mowed down the seedlings like tanks, despite the glass roof that was supposed to keep the diatomaceous earth dry. I guess it wasn't in the cards for us to have any carrots this year. Never before did we lose our direct-seeded fall planting of these two crops *completely* to slugs; it must be one hell of a great year for them.

The ratatouille, made of all our own ingredients—peppers, onions, tomatoes, seven big purple eggplants, thick tomato sauce—and also one small can of Contadina tomato paste—came out well. We ate some for dinner and froze seven more dinners' worth. Don't know how much more we can expect of these nightshade crops; the first frost can come any time now, there was a frost warning night before last. We're still eating the last cukes, picked on September 21, over two weeks ago! but the few left on the vine are yellowing toward ripeness and softness. I want to save the peppers for dicing and freezing—better keep some sheets at the ready, and an eye on the forecast.

The bird netting, held by nailheads around the perimeter of the frame, covers the whole pond except for the shelf end, where the tall plants—irises and cattails—stick up too high. We still have at least one frog, which may be glad of that foot of open space; I heard it plop in when I went to feed the fish this afternoon. The bees are bothered not at all by the netting and fly to and from the water clover as if it weren't even there.

October 9. This entry is a duck update.

The molt is finished and the mature plumage is gorgeous. Helen Stipe said at the open house that she once had a dress like that, shiny black, with purple and green iridescence. I keep trying to photograph them, but they won't hold still and the iridescent patterns shift with every movement. Probably the only way to get a good picture would be to make lots of exposures without trying for the perfect shot and hope one of them happened to capture the intense beauty of the colors. It's easy at last to tell ducks from drakes; the four males are now substantially larger, with thick green heads and necks much heavier than those of the two females.

And to our surprise, now that their adult plumage is in they can also fly, sort of. When we herd them from their pen to the garden, they flap their wings, run like crazy, and actually lift off for short distances. One flew about two or three feet straight up in the air the other day, when I was trying to get him out of the pen. They're not seriously going anywhere on the wing, that's for sure, but I'd thought "flightless" implied something less than even this.

And to prove that they've really grown up, yesterday I caught two of them in the bathtub *in flagrante delicto:* the first (witnessed) duck-fuck! I'd read that ducks prefer to mate in the water, the male seizing the female's neck in his beak and forcing her head under while climbing on her back to do the deed. That's what was happening, with the other four ducks clucking enthusiastically at tubside, necks bobbing up and down in a fashion somewhat reminiscent of *pigeons.* Clucking isn't quite what it is, but it's not quacking either—more like chuckling. Or cluckling. Ducklingcluckling?

October 10. An update to the update: Ted fed the ducks this morning and gave them clean water. When they mobbed the newly filled bucket, the big drake started with the neck-bobbing again, crowded up behind one of the ducks, grabbed her neck, and dunked her head *in the water bucket* while trying to mount her! One of the others broke it up. We both thought this hilarious—we now have not only a duckmuckbucket, but a duck*fuck*bucket.

I finished cleaning the dry beans—very laborious—and weighed them: a scant two pounds. They've been put to soak overnight.

October 19. Last night was the first frost. The ducks' garden bucket had ice in it this morning; the pepper and eggplant leaves are hanging like green rags. Autumn's well and truly here.

Some of the beehive frames stored in the back room were beginning to show signs of wax moth damage. In anticipation of the frost I put them out on the patio overnight. That one good freeze should have killed the moth eggs and larvae in the wax, so the frames of empty comb can be safely stored inside without further trouble.

Four days ago I took off the topmost box. It's a two-story hive again at last; the bees end the season as they began it.

October 21. This evening we emptied out the freezer and inventoried its contents, which are as follows:

spaghetti sauce (tomatoes, peppers, onions)	72 pints
tomato juice	17 pints
thick tomato sauce	3 pints
Romano beans	4 pints
mixed beans	2½ pints
wax beans	7 pints
Kentucky Wonders	2 pints
green peppers	1 gallon bag of diced pieces
jalapeños	1 quart bag of slices
sweet corn	2 bags, about 3 pints
applesauce	10 pints
eggplant ratatouille (eggplants, tomatoes, peppers, onions)	9 pints
bean paste (processed baked beans made from dry beans)	1½ pints

We've eaten tons of all these things already; the above is just what's left to take us into winter.

Here's the inventory of stored items other than frozen ones:

canned
> blueberry jam 7 pints (8 half-pints already consumed)
> dill pickles 5 quarts, 1 pint
> bread-and-butter pickles 3 quarts, 7 pints

stored
> onions one crisper drawer, between 15 and 20 pounds
> potatoes, all varieties 15 pounds
> sweet potatoes, all varieties 15 pounds
> popcorn 24 ears, assorted sizes
> dry beans some still drying on a tray, maybe half a pound

plus
> green tomatoes several dozen
> eggplants 5 small ones, to be ratatouillefied
> honey 50 pounds

October 25. The flock is down to four members. We butchered two more ducks this week. It doesn't get easier—or the killing doesn't; the skinning did go faster yesterday.

The book's right: ducks don't gain weight after eight weeks, for all that the drakes at least look so much larger in their new plumage. Our first, second, *and* third carcasses, dressed out, each weighed just under three pounds. The new apparent weight is all fat and feathers, great mats of both; no wonder ducks are essentially weatherproof and able to stay outdoors all winter.

The drakes are too long now to hang from the gallows over the compost bin without dragging on the ground. Ted tied a rope between the hammock post and the hook in the pin oak, and we hung the ducks from that. Their behavior seems odd but makes evolutionary sense: they *hate* being chased, squawk and run frantically to get away, but once caught they sit rather calmly on my arm, look around and don't vocalize. I was terribly aware of the first duck's high body temperature as I carried her to her fate; these two drakes were not so warm, both because I was wearing double sleeves, and because *they* were wearing so much fat-and-feather insulation. Even when we tied their feet and suspended them they didn't say anything, and only one even flapped his wings much. It's as if, once caught, they accept their fate. (The second one was hard to stun—all those feathers—which made the process tougher on all concerned.)

Ted was reluctant to grab their bills—kept flinching away. Then once grabbed he would *lift* them by the bills to saw through the feathers to the jugular vein, instead of stretching the necks out to create a tension the knife could cut into more easily. I slit the second one's throat myself. For me to take a turn was only fair. Ted doesn't like doing this any better than I do, and gets rattled. I must have cut a nerve as well, one of the nerves severed by *chopping* off the head, because although the duck was bled out and obviously dead, it twitched and shuddered its beak and lifted its head sideways several times, which generally prolonged the awfulness.

But we did it, and skinned and cleaned them, and both are now in the freezer, awaiting a time when we can do justice to the dinners they'll provide. I'm heartily thankful there's only one more left to butcher (for Thanksgiving) before winter comes.

I reflect again that this wouldn't be so difficult if ducks were part of our regular operation year in, year out, a new generation always coming on to replace the old one; but we started with seven—a finite number—and working our way down to three has seemed anything but routine.

I've been reading *Loving and Leaving the Good Life*, Helen Nearing's account of her life with Scott, and find to my surprise that in many ways Scott Nearing sounds insufferable: self-righteous, critical, a *terrible* father. Helen may be parroting him when she talks about why they were both vegetarians, going on about how we share the world with many fellow creatures and have no rights over them whatever. She and Scott disapproved of pets and working animals too. Neither of them seems to have thought very deeply about the way our ecosystem operates: that everything lives by eating something else, that life subsists by killing and devouring life. She draws a categorical, but automatic, distinction between animal life and vegetable life—a mistake that Sue Hubbell, by contrast, is scrupulously careful not to make—and sees no evil in taking the lives of lettuces and carrots. Yet the whole subject is fraught with a rather Thoreauvian self-righteousness: *If you don't do as we do, you don't do as well.*

All this gives me a pain. Harlan's and Anna's attitude seems far more sensible; they kept dogs and bees, raised and slaughtered goats, set out their trotlines, thankfully ate the groundhogs their big hound Ranger used to catch. We're part of it all, not above it all. I don't like the system either, as I've said repeatedly, but this experimental year has *not* been about denying the true and actual nature of things.

The four ducks got into the outer pen and somehow managed to knock the plug out of their bathtub drain last Sunday night. I went out Monday to feed them and discovered I'd forgotten to close up the inner pen after working in there Sunday afternoon, and that the tub was empty of everything but a little mud. I refilled it, but whatever happened seems to have put them off swimming because they haven't been back in the water all week. Today I mucked only, no need to siphon— the water isn't dirty. Full of white-oak leaves, though, which I scooped out and threw over the fence.

I gave the bees their first dose (of three) of Terramycin in powdered sugar this week. When that's finished I'll make some sugar syrup and dose them with Fumadil. Then I'll make them a grease patty (a *Bee Gleanings*–recommended alternative to menthol), wrap the hive in roofing paper, and call it a season.

The garden has taken on an agreeably autumnal aspect, in which the duck survivors look just right, snoozing in the leaves or rooting around for bugs and slugs. I picked the beans from the last three poles and shelled them out. All that's left alive out there are the brussels sprouts, a few of which I harvested and froze this week, and the snow peas. Some of the latter are very vigorously climbing their brush supports, but none show signs of flowering anytime soon. Oh well. If they make it, great. If not, it was worth trying.

Apart from moving the ducks permanently into the garden, the last remaining major seasonal chore is to get the pond ready for winter.

November

November 11. It's been a bloody cold week. Today, though it only got up in the middle fifties, was still my best chance before Thanksgiving to winterize the pond; so I rolled up my sleeves and tackled it. Brrr. The water is *much* colder than 55°; it's a wonder the fish can still swim around. Not a pleasant job in other ways, either. I peeled back the netting and began by pulling out the remaining water hyacinths and dumping them in a ten-gallon bucket. They at least were floating, so I didn't have to stick my hands in the water. But when all the hyacinths were out, and the irises and cattails had been crew-cut, there was no help for it. By the time I'd scooped all the oak leaves out, and severed all the lily-pad stems close to the soil line, my right arm was beet-red to the elbow and I could hardly work (or feel) my fingers. The water lilies were easier at that than the water clover, which had grown in a tangled mat, hard to shear with the dull scissors.

Almost everything was moribund if not dead, except the one vigorous Comanche water lily; it had some young red leaves and was actually sending up a last die-hard bud, which I amputated with regret. I pulled the water-clover container off its bricks and let it settle on the bottom, ditto the floating-heart container off its crate, which immediately floated to the surface and got removed.

When I was shearing away at the water clover, a disoriented, uncoordinated green frog darted out right under my nose and tried to hide between the liner and the board I was kneeling on. It had obviously been hibernating peacefully in the water-clover foliage—not burrowed into the mud, not yet anyway—and been disturbed by my activities. I'd wondered whether any of the frogs were still around. I saw it give up on the liner and swim into an algae-covered oxygenating plant, and I never saw it reappear, so I guess it resumed hibernating in the new location. Hope so; hope we have frogs and tadpoles galore in the spring.

I'd intended to cut the submerged plants back too, but the water was just too cold to get into. Anyway, that frog needs somewhere to sleep. And maybe there are other frogs, and bullfrog tadpoles too. Following the Lilypons instructions, I set all the bog plants on the bottom of the pond (four against the side nearest the house, there being too little space at the foot of the shelf for them all), skimmed off all the floating debris and pine needles I could, and replaced the net. We're going to need more than netting if the weather goes on being this cold. We mustn't let the pond freeze over, and electrically powered means of preventing that are out. The pond looks as awful right now as it did while it was curing, and turned that murky green color. The fish seemed dazed, with most of their cover suddenly gone.

Other activities since the previous entry: on October 30 I moved all the robust-looking water hyacinths inside. They're sitting in the dining room by the patio door, in the washtub. Being tropical, water hyacinths can't cope with even this much cold. I like having them there, actually. Day before yesterday I bought eight goldfish ("large feeders") for a dollar, and released them into the washtub after floating their plastic bag in it for a couple of hours. The fish had been very active in the pet store, but they fled gratefully into that mate of water-hyacinth roots and became almost motionless. They wouldn't eat at first, having been kept in "goldfish keeper" that turns the water greenish blue (and suppresses appetite, I guess), but by dinnertime yesterday the keeper had worn off and they were at the surface chomping on the goldfish food. That's the ticket. If they don't eat, they don't poop, and the water hyacinths don't keep healthy. Eight may not be enough; I'll get some more if these do well.

I've finished the three Terramycin treatments, working without gloves or veil in weather so chilly the bees were too torpid to fly. And I bought five pounds of sugar to make syrup. Maybe Friday. We took down the bee barrier last weekend, cut it loose from the laths in the center and rolled it up to store away for a future conversion back to its true purpose of shade netting. Nice not to have the view blocked in that direction, but what an excellent solution to the problem that barrier proved to be!

The garden has virtually shut down except for the pea vines, several of which are trying to flower. Even if they do, the bees aren't flying; it's too cold. So I don't suppose we'll get any snow peas anyway, unless things warm up (and that's not what the forecast says). Colder than normal this fall, just like the spring and summer. The brussels sprout plants are still alive, but the temperature went down to 25° Sunday night so Monday I harvested all the sprouts left on the stems. Some were marble-size and the rest went down from there. We had some for dinner that night: a little strongly flavored, maybe, but great mixed in with other things. We'll work our way through the rest before I leave for the Kentucky Book Fair next week; not enough to bother freezing, probably.

The ducks go into the garden for several hours every day, and bustle about or snooze in the leaves, until they decide it's time to go home, and start quacking. Their pen smells unspeakably bad, but the big permanent move to the garden, house and all, is planned for Thanksgiving break, only two weeks away, and I guess we can all stand it till then. Apparently the cold affected the rubber stopper, because after a week of standing there, one morning all the water had drained out of the tub again, with no sign the ducks had been anywhere near it. Their water bucket in the garden had frozen over too, so I'd say it's time to disconnect the hoses and store them, and switch to the more bothersome daily "bucket of warm water" routine recommended in one of my books. *Months* of lugging a bucket of water to the garden every morning . . . (you wanted these ducks, remember). Monday I bought them another fifty pounds of feed.

November 13. Three of the eight goldfish died of some malaise they all seemed to have, but the other five have recovered and are behaving normally. They seem to appreciate the little clear space I created in the water hyacinths, and tend to foregather there.

This afternoon Liz brought over the worm farm Jeff had been given at GardenFest!, teeming with red wigglers that had pretty much consumed their newspaper bedding. He's through with it, so now I've got worms. Too late for fish feeding this year, but next spring I'll be all set. The worm farm is a plastic box with two perforated tubes (for air) running through it and drainage holes in the bottom. I moved the contents to one side and added more bedding (strips of torn newspaper) and some nice garbage to the other side. In theory they'll eventually move over into the new bedding and I can take out the castings, etc. and put them in my houseplants or whatever. Use them in the potting soil for starting seeds. There are always worms right at the rim and on the underside of the lid when I remove it. Some instantly try to escape over the top. Suggestive of what may have happened in my bins, out in the yard.

November 16. Yesterday morning, following several below-freezing nights, I officially gave up on the peas and let the ducks in to chomp on them, which they were glad to do; also the brussels sprouts; also the slugs they find in the leaves we rake up in the yard and dump in the garden.

I drained and stored the hoses this afternoon, before they could freeze some fine night and burst. The bucket brigade begins tomorrow. In the hyacinth washtub, another dead goldfish.

November 26. Thanksgiving. We celebrated the holiday with an almost-100-percent homestead dinner. Duck, sweet-potato casserole, brussels sprouts, applesauce, gravy and stuffing made with neck-and-giblet stock. We roasted the duck under aluminum foil (it being skinless) and the meat was fine-grained and not at all dry—but gray-colored and extremely *lean.* Both of us prefer it in the form of leftovers, cut up in gravy, over stuffing. The stock is superb, so I think we'll cut the rest of the ducks in pieces and boil them—instead of roasting or whatever—to get as much stock as we can.

Two more dead goldfish. Vicki, veteran of many pet goldfish

deaths, says they never do seem to live very long. I think it must be more a question of providing the right environment, or protecting them from the wrong one, and that in some fashion this one is wrong.

November 28. Yesterday we moved the ducks and their house and bucket into the garden for the winter. Potentially it's a scheme that should benefit everybody: the ducks, because they'll have lots more room (and can be safely shut up in their house at night), and us, because we lost no time in taking apart the pens and cleaning up that smelly area behind the garage. The only problem with the plan so far is that all attempts to herd the ducks into their house have been unsuccessful; they can't seem to recognize it in the new setting and refuse to go in. They make it plain that what they want is to go "home" to their pen—which, of course, isn't there any longer. I'm thinking of devising some sort of chute to funnel them into the house with when I get some time; meanwhile it's upsetting to see their obvious distress. Let's hope they adjust in another couple of days.

We raked leaves all morning, then spent this afternoon turning the garage back into a garage, after its long season as a "sort of barn."

November 29. We're relieved and encouraged to see the ducks working their way nearer to their house, where their water and feed are. All may yet be well.

Yesterday and today I made 2:1 syrup for the bees (ten pounds of sugar, five pints of water), mixed in the Fumadil, and started feeding. And Ted and I finished another project: getting the pond covered so it won't freeze. Across the pond's waist we laid a four-by-four-by-eight-foot beam we happened to have around, and across the beam a three-foot-wide length of chicken-wire from the disassembled duck pen. The wire stretched from end to end. Over that went my old nine-by-twelve plastic tarp, tacked down all around the edges, with just the right amount of space left uncovered at the deep end. The beam-and-chicken-wire arrangement holds the tarp nicely above the water. Finally, to keep the tarp from blowing, we replaced the bird netting;

hooked over the nails, it holds the whole arrangement snugly in place. That should do the trick. The tarp won't rip out, and the pond won't freeze, and the fish and plants will make it through the winter. Since its crew cut, by the way, that one die-hard Comanche has set up two new leaves and a flower bud.

November 30. This morning Ted came in at 7:15 and said, "Sorry to wake you, but there's been a calamity with the ducks."

While I staggered up and started pulling on boots and a coat, he explained. He'd finished his breakfast, and was about to go out and feed the ducks theirs, when he looked out and saw a duck right on the path to the patio. Thinking one of us must have left the garden gate open, he rushed out to see what was what. The gate was shut, but one duck had disappeared altogether. Of the three remaining, one was hurt and flopping around on the ground, and another—the one he'd seen— had flown over the fence to escape the predator, whatever it was: fox or raccoon, probably. Or coyote, if they've made it into the Philadelphia suburbs.

I was very short of sleep, and to be confronted in that state with the evidence of a mistake so dire in its consequences was frightful. But we had to spring into action at once, because Ted had to leave in almost no time for his 9:00 class.

I got the cleaver and we went out to the garden. The injured duck had crippled over and was trying to drink out of the bucket from its pathetic, broken stance on the ground. Ted fetched a round log of firewood. I caught the duck, and I held it while he chopped off its head with one decisive stroke. The instant he did so I let go of the body, thinking it would struggle and get blood all over me; but the shock must have affected its nervous system some way. Blood sprayed in a fine stream from the stump, as if under pressure, but the body didn't so much as twitch.

Very soon there was no more blood. Ted had already put the loose duck back into the garden. He jumped in the shower; I put the carcass in the sink and plugged in the kettle.

After breakfast I tackled the skinning and butchering, the former immensely more difficult to do without somebody else to hold while I pulled. All the time I worked, in that mildly awful stench of blood and entrails, I felt just terrible. When had it happened? Had the poor thing suffered for hours? Why, oh why, would none of them go near their beautifully retrofitted house to be shut in safely for the night? Why had we ever let ourselves believe the electric fence would keep out predators? From habit we'd kept on shutting the ducks up in the roofed-over inner pen every evening, but for a good while now that had really seemed like overkill. Obviously, it hadn't been. Equally obviously, the two survivors would have to be caught and shut in their house until they learned to go into it of their own free will.

When I finished butchering and had cleaned up the mess, I went back and walked around the garden and then the fence line. No sign of entry: not dog hairs snagged on the wire, not a hole dug under the fence, not a stray feather, let alone any grisly remains. I dumped the bucket and brought clean water. Then I left the pair of survivors alone all day.

At dusk I went out to catch them and shut them in their house. As soon as I cornered the female against the fence, she simply flew over it, demonstrating how she'd escaped the danger last night. I herded her back inside. Then I spent half an hour chasing the poor nitwits around the garden, trying first to catch them, then to herd them into their house, then—a last desperate measure—to throw a sheet over them; but I only succeeded in freaking them out so totally that they escaped through the back of one of the compost bins and hid under the big rhododendron by the garage. (A rabbit shot out as they shot in.) Meanwhile it was getting darker and darker. I went inside to let them calm down. When I came back out they were standing, completely disoriented, on the site of their former pen—the place they stand and quack and yearn for every day at dusk in the corner of the garden—but there was no pen, no house, nothing but the bathtub with some oak-leaf-stained water in the bottom.

After a lot of effort I finally got them back inside the garden.

But by then they were so spooked there was no earthly way I was going to catch them; and there we left it.

Oh boy. Enormous error, to try to change their habits and expectations at this late date. The predator will certainly be back for more duck. The female can probably take care of herself (unless he catches her asleep), but I'm afraid the drake is doomed. I wish like anything we'd left well enough alone, stench and mess and all. A lesson learned the hard way. For all that I tell myself over and over that this, too, is part of the homesteading experience, and that a duck's whole purpose in life is to be dinner for some predator, it still feels just awful. Ted claims to be haunted by the Beatrix Potter story of Jemima Puddleduck and the fox; as for me, *all day long* that song "Fox Went Out for a Feast One Night" has been running wretchedly through my head:

> He grabbed the gray goose by the neck,
> Flung a duck across his back,
> He didna' mind wi' the quack-quack-quack
> An' the legs all danglin' down-o.

December

December 2. Duck update: yesterday, while I was in town teaching, Ted reinforced the empty compost bin with chicken wire and rigged a window screen across it for a door, with hinges made of pieces of a web belt. Then he spent half an hour herding the spooked ducks into it and closing them in. When he whooshed the door shut, they went nuts: bashed themselves against the wire and quacked hysterically: *Trapped! Trapped!* But when I went to let them out this morning, all was serene. They stayed right where they were for several hours despite the wide-open door, before venturing forth. I looked out the window around 10:00 and saw them snoozing in the cornstalks, and later they'd been foraging for slugs (sticky bills).

Around 3:00 this afternoon, I went out and began herding them back to that corner of the garden. Right away the female started with some behavior Ted had mentioned yesterday: twisting her head around in a ducking, preening gesture, an obvious nervous tic. She was also making the food-gathering sound, inappropriate in the situation. The poor thing is still as freaked as she can be. I recall my sister's little dog, Ginger, who got so frightened once during a bad thunderstorm and was never the same afterwards.

It only took me fifteen very patient minutes to get them into the bin today, but several times when I almost had them in they broke and ran, and I had to start over. It'll be *much* easier with two of us. The trick is to edge them ever so carefully where you want them to go, an inch at a time. I'm hoping that if we just go on with this unvarying routine, they'll gradually adjust and let themselves be herded like they used to.

After this, will the female lay? Ever? I just wonder.

December 4. Spoke too soon. It *was* easier yesterday, when Ted and I put them to bed together. But today they stayed in the bin all day long until I went out to tuck them up, whereupon they came out and flew right over the garden fence—not just the female, but both of them! They haven't been eating this week. I guess even the drake had slimmed down enough to become airborne. As for the duck, she's so thin she was able not just to clear the fence but to fly the whole way over to the garage. I followed on foot. By the time I arrived the drake had joined her, and they stood quacking sadly in the houseless and fenceless site of their destroyed home.

I herded them back into the garden—the conditioning still serves for that, they file through the "gateway" of the phantom pen even though there's no fence anymore and head back to the garden gate—then had the devil's own time getting them into the bin. The female kept twisting her head the whole while. God almighty! Who could have guessed they'd be so hard to accustom to a life *we* thought would be more enjoyable for them?! This is one well-meant error none of us may recover from. They might have done better in a totally different setting than with a different routine in the old one.

Third bee feeding today. A couple more, and it'll be grease-patty time.

December 7. Ted let the ducks out of their bin this morning, so I didn't know last night's wind had blown the fiberglass panels off the top till I had already herded the ducks (with great difficulty) back into it for the night. It had been so hard to get them to go in that I didn't like to let them out again while I put the roof back on, so I went ahead and did it while they were shut inside the bin. But when the panels rattled the ducks went crazy, flapping against the wire, trying to get out. I put the heavy old wooden ladder on top to keep the lid on, windy nights or no. But I also took a good look at the ducks. They've lost weight appallingly. No wonder: they don't eat. Or drink: today, again, they didn't touch their water. They did come out, but spent the whole day either craning at their former home or crouching against the fence, as close to it as they could get. When I was coaxing them back into the bin, I noticed the drake was trembling all over—from cold? from nerves? They've lost so much body fat, I wonder if they mightn't feel the cold.

Saturday, day before yesterday, they seemed better—played around, nibbled frozen brussels sprouts, snoozed under the forsythia fence, actually *flew* back and forth inside the garden, getting about six feet of lift—and I was encouraged and hopeful that they were starting to settle down. But then it got very windy, and I guess the wind rattling the fiberglass rattled their nerves as well. Yesterday they stayed in the bin all day, never came out at all. Then last night the lid actually blew off. Tonight they absolutely did not want to go in. Every scare undoes whatever good's been done and probably increases the net harm.

So, *very* reluctantly, I faced up to it. I'm going to have to move their house back and rebuild the old inner pen. It may not do the trick— they're so badly freaked out, maybe nothing will do any good at this point—but it's all I can think of to try, and I have to try something. If Ted will help me move the house, I think I can do the rest myself. Maybe Thursday, if they can wait that long.

December 10. This morning Ted and I moved the duck house back to its old place behind the garage, and put up and roofed the old pen around it. It took the two of us from 10:00 to noon to accomplish this task, working through the second hour in a light, freezing rain.

At noon we finished and went to get the ducks, who don't seem to have moved in about three days from their position against the back of the bin—though there's been new poop all along to prove that isn't so, and feed does disappear, if slowly and mostly at night. I drove them out with a wand. The female almost took off flying and we both froze till she gave up that idea. We herded them carefully toward their reconstructed home. If we were hoping (and we were) that they'd run toward it quacking with joy, we were in for a disappointment; it's four or five days too late for that. But—after a feint toward the big rhododendron—they did go into the pen. I had carried their water bucket over as I came, and as soon as I put it down on the screen the drake and then the duck started drinking thirstily. Greatly heartened, we shut the gate and went away.

All afternoon they've been standing there by the bucket. I went out once to check on them and found both standing on one foot, heads tucked down, evidently sleeping—though they woke up when they heard me coming and clucked and twitched around nervously. No sign of any interest in their house, which now looks exactly as it did before. This resembles catatonia, the consequence of extreme fear; but maybe in time they'll rally if things go on calmly from now on.

It's an immense relief to have the ducks back where they so manifestly wanted to be. I only wish I'd realized sooner that it would come to this in the end, and acted before the final round of terror reduced them to this pitiful state.

I've learned something from all this: the difference between raising livestock to be butchered quickly after a happy life, and keeping livestock in a state of misery unintentionally caused by you.

I looked into the beehive, but they still haven't finished the last jar of syrup. Down the hatch, guys. I want to wrap the hive and be done.

New Year's Eve. During the past three weeks the ducks have come a long way back from the scrawny, psychotic creatures they'd turned into have put on weight and started behaving more normally (or less abnormally)—playing in the downspout mud, dabbling in the bucket, muddying up their water, and pecking away at their feed as soon as we toss it in. But they'll never be the confident, quacky ducks of yore; their innocence is gone, they know it's a tragic world.

The drake looks like his old self and can no longer fly over the garden fence. The female has plumped up too. But they don't like going into the garden any more *at all*, balk at the gate, do nothing once inside but walk up and down along the fence watching the door of the house for me to come out and release them, and if they have to wait very long the female flies over the fence. I hear quacking and carrying on, and there she is out in the yard, and there the drake is inside the fence, having a fit. I herd her back inside easily enough—they hate being separated now, they're all in all to each other—but the point is, they won't play in the garden anymore. It represents danger and they hate going in.

We keep putting them in anyway, in hopes that they'll get used to it, but it's terribly hard for just one person to get them out of their pen. They run from one end to the other, back and forth, and then the female starts that chuckling-and-head-twisting behavior, and I start feeling bad about tormenting her further after all she's been through. It must have been a different duck, not either of these two, who used to lead the flock through the gate.

But their behavior *inside* the pen has pretty much gone back to what it was before the catastrophe. The night of the day we moved them back, December 10, brought fifty-mile-an-hour winds that blew the fiberglass panels, ladder and all, right off the compost bins; one even landed on the other side of the garden fence. If we hadn't gotten them out of there when we did, I think we'd have had to butcher them; they'd never have gotten over the trauma. But we did get them out. Until now I hadn't realized how well their pen was located and appointed, a case of not realizing you've been doing something right until you mess up trying to improve it.

On the second day, a rainy, cold day, I happened to be watching when they finally ventured into their house. The drake stood for a long time outside it, peering in, hesitating. Then, one step at a time, he went in, and the female followed. They'd been playing in mud over by the downspout for a good while, and had eaten their feed when I brought it—though not till I'd withdrawn a fair distance—and all that was encouraging; but when they went into the house I let myself start to believe they'd be okay, and I guess they pretty much are.

Betsy Stedje, two doors down, says she's seen a fox in her back yard, so I guess we know who killed Cock Robin.

The fifty-mile gales ripped the grommets right out of the old plastic tarp on the pond. We took it off and put on a brand-new one, fiber-reinforced, with rope around the edges.

For the past couple of weeks the weather's been either very cold or warmer but rainy, and I haven't been able to open the hive. But yesterday the rain stopped for a while, long enough for me to make a grease patty—two parts sugar to one part Crisco—and push it down on top of the frames, under the inner cover. I probably pushed it down on some bees as well; they were thick on the top bars and sluggish, and I didn't want to leave them uncovered any longer than I had to. But that final chore has at last been done; and today was so warm they were flying, so they must have been able to break their cluster and chomp away on their grease patty, too. It's very late, but maybe not absolutely too late, for this treatment to ward off the tracheal mites in whatever mysterious way it does so.

I took off the empty hive body that had been enclosing the feeder jar, removed the empty jar, brushed the dead bees off the inner cover, and replaced the telescoping cover. Then I wrapped the whole thing up in the same length of roofing paper I used last year and tied it with rope.

And that's that. Hive, pond, ducks, all buttoned up for winter.

Yesterday I brought out the last dozen green tomatoes to finish ripening on the kitchen windowsill. We're working our way through the put-by food, but there's lots left in the freezer and basement, including the last duck, the one injured by the fox. Christmas dinner was a repeat of Thanksgiving, except we didn't roast the duck. We boiled it, cut it up in gravy made from the stock, and served it spooned over stuffing made with more stock. Superb. And for dessert, two "pumpkin" pies made from sweet potatoes.

The Homestead Year is finished. But the worms are munching on garbage in the bedroom; the fish are waiting under tarp and chicken wire for spring, the ducks for egg-laying season to come round, and the black currants to produce enough berries for jam. So some of what we started carries on, into the new year.

Epilogue

Nineteen ninety-three turned out to be an unofficial extension to the Homestead Year. The weather was kinder and the garden produced bumper crops of tomatoes and sweet corn. Two apple trees, Prima and Liberty, drowned us in applesauce, though neither type proved to be a good fresh keeper.

Intimidated by my difficulties in locating the queen, I never did provide my wild hive with a tame queen bred for gentleness and reluctance to swarm. To the best of my knowledge, however, no swarms occurred. We took forty pounds or so of honey off in September— a smaller harvest because a later one; by September the bees had eaten part of what they'd stored in the supers. I was stung a good deal, but both bees and beekeeper had a less traumatic year.

Using an antique taper mold contributed to the project by Liz Ball, we finally made ten beautiful, smokeless, dripless, pure beeswax candles out of two seasons' worth of cappings wax and some purified scraps.

Tarp notwithstanding, the pond froze solid on top and stayed frozen throughout the entire month of February. Happily, plants and fish came through unscathed. We fed redworms aggressively, and at summer's end broiled and ate five of the bluegills and one catfish. The bluegills were little and bony but delicious, the catfish much larger— big enough in fact to look respectable on a plate—and also very tasty. My recommendation to anyone interested in setting up an edible pond: stock catfish only. (It's what I'll do in future.)

The duck saga continued on its riveting course. Both ducks recuperated fairly well, emerged from winter in good shape, and mated vigorously in the spring. Since there was no question of butchering

these two, Ted gave them names: Winnie and Sir Francis, soon short-
ened to Frank. After laying a delicious daily taupe-colored egg for some
weeks, Winnie "went broody," built a straw nest under her house, and
hatched out three coal-black ducklings. Only one survived her rough-
and-ready mothering. We called the survivor Little, a name that at least
had the virtue of being androgynous. (He turned out to be male.)

Little had a terrible childhood. Frank hated him on sight and
tried his best to kill him. This completely unforeseen development created enormous hassle. For three months we had to keep a fence between
Frank and Little, or Frank would go for him. Some of Winnie's time was
spent on Little's side of the fence, some on Frank's, and each com-
plained and moped when Winnie was with the other. Very soon Winnie lost all interest in her son; we made her keep Little company some
of the time, because he was so unhappy without her, but she would ignore him and snuggle with Frank through the chicken wire. We were
constantly separating and recombining frustrated ducks. All attempts
to get Frank to tolerate Little failed. It was a distressing, if preposterous, situation.

I'm extremely happy to be able to say that at twelve weeks Little was adopted out to a duck-raising farm family in South Jersey, and
that at last report he had settled in well, and had a girlfriend.

I still don't know why we haven't got any earthworms.

Bibliography

Aebi, Ormond, and Harry Aebi. *The Art and Adventure of Beekeeping.* Emmaus, PA: Rodale Press, 1975.

Appelhof, Mary. *Worms Eat My Garbage.* Kalamazoo, MI: Flower Press, 1982.

Ball, Jeff. *The Self-Sufficient Suburban Garden.* Emmaus, PA: Rodale Press, 1983.

Bartholomew, Mel. *Square Foot Gardening.* Emmaus, PA: Rodale Press, 1981.

Belanger, Jerome D. *The Homesteader's Handbook to Raising Small Livestock.* Emmaus, PA: Rodale Press, 1974.

Bilderbeck, Diane E., and Dorothy Hinshaw Patent. *Backyard Fruits and Berries.* Emmaus, PA: Rodale Press, 1984.

Bonney, Richard E. *Hive Management.* Pownal, VT: Storey Communications, 1977. (A Garden Way Publishing Book.)

Holderread, Dave. *Raising the Home Duck Flock.* Pownal, VT: Storey Communications, 1978. (A Garden Way Publishing Book.)

Hostetler, John A. *Amish Society.* Baltimore: The Johns Hopkins University Press, 1963.

Hubbard, Harlan. *Payne Hollow: Life On the Fringe of Society.* New York: Eakins Press, 1974.

Hubbell, Sue. *A Book of Bees.* New York: Ballantine Books, 1988.

Lawrence, James M., ed. *The Harrowsmith Country Life Reader.* Charlotte, VT: Camden House Publishing, 1990.

Logadon, Gene. *Getting Food From Water.* Emmaus, PA: Rodale Press, 1978.

————. *Homesteading: How to Find New Independence on the Land.* Emmaus, PA: Rodale Press, 1973.

Luttman, Rick, and Gail Luttman. *Ducks and Geese in Your Backyard.* Emmaus, PA: Rodale Press, 1978.

Nearing, Scott, and Helen Nearing. *The Maple Sugar Book.* New York: John Day, 1950.

Perrin, Noel. *Amateur Sugar Maker.* University Press of New England, 1972.

Robinson, Ed, and Carolyn Robinson. *The "Have-More" Plan.* Pownal, VT: Storey Communications, 1973. (A Garden Way Publishing Book.)

Rodale's All-New Encyclopedia of Organic Gardening. Emmaus, PA: Rodale Press, 1992.

Simonton, O. Carl, M.D., Stephenie Matthews-Simonton, and James L. Creighton. *Getting Well Again: A Step-by-Step, Self-Help Guide to Overcoming Cancer for Patients and Their Families.* New York: Bantam Books, 1978.

Thomas, Charles B. *Water Gardens for Plants and Fish.* Neptune City, NJ: T.F.H. Publications, 1988.

Thompkins, Enoch, and Roger M. Griffeth. *Practical Beekeeping.* Pownal, VT: Storey Communications, 1977. (A Garden Way Publishing Book.)

Vivian, John. *Keeping Bees.* Emmaus, PA: Rodale Press, 1986.

———. *The Manual of Practical Homesteading.* Emmaus, PA: Rodale Press, 1975.

———. *Raising Ducks and Geese.* Pownal, VT: Storey Communications, 1977. (Garden Way Publishing Bulletin A-18.)

von Frisch, Karl. *Bees: Their Vision, Chemical Senses, and Language.* Ithaca: Cornell University Press, 1950, 1971. Revised edition.